View of a hamlet in the morning.

Peasants, Subsistence Ecology, and Development in the Highlands of Papua New Guinea

by Lawrence S. Grossman

PRINCETON UNIVERSITY PRESS
PRINCETON, NEW JERSEY

Copyright © 1984 by Princeton University Press

Published by Princeton University Press, 41 William Street,
Princeton, New Jersey 08540
In the United Kingdom: Princeton University Press,
Guildford, Surrey

All Rights Reserved

Library of Congress Cataloging in Publication Data will be
found on the last printed page of this book
ISBN-0-691-09406-3

Publication of this book has been aided by the Whitney Darrow Fund of
Princeton University Press

This book has been composed in Linotron Trump

Clothbound editions of Princeton University Press books
are printed on acid-free paper, and binding materials are
chosen for strength and durability
Printed in the United States of America by Princeton University Press
Princeton, New Jersey

TO MY PARENTS HERMAN AND FLORENCE

CONTENTS

LIST OF FIGURES

LIST OF TABLES

CONVENTIONS

ABBREVIATIONS

DASF	Department of Agriculture, Stock and Fisheries
DPI	Department of Primary Industry
IBRD	International Bank for Reconstruction and Development
IDA	International Development Association
PMV	Public Motor Vehicle
PNGDB	Papua New Guinea Development Bank
PWD	Public Works Department

CURRENCY

Before 1975, Australian currency was used in Papua New Guinea. In 1975, Papua New Guinea established its own currency, the Kina. One Kina equals 100 toea. At the beginning of fieldwork in April 1976, the Kina and the Australian dollar were approximately equivalent in value. In November 1976, Australia devalued the dollar to $A1.00 = K0.87. The U.S. dollar was worth K0.79 in 1977 and K0.69 in 1981.

ORTHOGRAPHY

A glottal stop is indicated by ('). The vowels used in spelling Tairora words have the following English equivalents:

a	as in *a*lone
aa	as in f*a*ther
ae	as in b*a*y
ai	as in *i*ce
au	as in c*ow*
i	as in te*a*
o	as in h*oe*
u	as in fl*u*te

NOMENCLATURE

After independence in September 1975, the names of government departments and administrative units were changed. Relevant changes are:

pre-September 1975	*current*
District	Province
Sub-district	District
Department of Agriculture, Stock and Fisheries	Department of Primary Industry

In referring to the governing authority existing in Papua New Guinea before December 1973, the term "Australian administration" is used, and for after this date "government" is employed. Papua New Guinea was granted self-government in December 1973.

FOREIGN WORDS

Many foreign words and phrases are either in Tairora, the traditional local language, or in *Tok Pisin*, the *lingua franca* of much of the northern half of Papua New Guinea. To distinguish between the two, I usually specify in the text which terms and expressions are in *Tok Pisin*. Others, except in obvious cases, are in the Tairora language.

Throughout the Third World, governments, planners, and international development agencies all view the expansion of rural commodity production as a fundamental prerequisite for national development. They exhort members of rural communities to increase their involvement in commercial activities, promising not only that such a commitment is for the good of the nation, but also that it will improve their welfare, provide security, and bring "progress" to their community. Contrary to such assurances, however, rural involvement in commodity production actually can undermine the viability and resilience of local communities. This book, set in the Highlands of Papua New Guinea, is a village-level case study of the dramatic impact of commodity production and the commercial economy on Third World rural communities.

The study focuses on the far-reaching implications of the conflicts between subsistence and commodity production using a cultural-ecological perspective. The main portion of the book deals with the eager and enthusiastic peasant involvement in commercial cattle raising and coffee production that occurred in 1976 and 1977. The effects of these activities reverberated throughout the local system: subsistence production was radically undermined, the environment seriously degraded, and inequalities in the control of wealth and resources substantially increased. The nature and impact of this intensive involvement and the causes of the subsequent decline in commodity production in the early 1980s have significant implications for understanding the development process in rural areas.

My approach departs significantly from most research on Third World development. I employ a cultural-ecological perspective, which focuses primarily on the numerous dynamic forces—both technical and social—affecting patterns of resource use and production. Thus, my analysis ranges widely to include many factors—the natural environment, subsistence and commodity production, social relations of production, cultural values, and political-economic forces—portraying the systemic relations among them in the process of change. This holistic orientation is essential for revealing the complex linkages involved in the conflicts between subsistence and commodity production that undermine the welfare of rural communities. Other approaches to the study of rural development often are less encompassing, with resulting disadvantages. Some underestimate subsistence-commodity production conflicts because they focus narrowly on isolated factors of production, thus failing to uncover the crucial linkages in the local system that are revealed by a more holistic approach. Others that do acknowledge that the commercial economy has in some way disrupted subsistence production often do not specify the exact mechanisms involved in the changes. In addition, many studies fail to appreciate the significance of the environmental context in which production takes place.

Furthermore, I challenge the conventional wisdom, which calls for radically increased rural commodity production to spur national development. Instead, I assert that priority must be given most urgently to the maintenance of subsistence production to ensure village welfare. When subsistence production is radically undermined by the commercial economy—an all too frequent occurrence—villagers often suffer dire consequences.

I stress in particular the importance of classic economic "boom and bust" cycles and their effects at the local level. The analysis of periods of both intensive commodity production and reduced commercial activity provides insights into the fundamental, perennial problems of the slow rate of change in rural areas and the inability of socioeconomic processes to radically transform peasant societies. To help understand why peasantries in different parts of the Third World follow divergent paths in response to boom and bust cycles, I present a model based on the degree of local autonomy, the nature of political-economic forces, and the risks involved in expanding commodity production. This longer-term view considering boom and bust cycles is essential for understanding patterns of change in rural areas. Other studies usually have a more limited time frame: some focus on the negative effects of the commercial economy only at a particular point in time; others emphasize only the constraints to expanding cash cropping. Both fail to consider the significance of oscillations in rural commercial activity over time.

Throughout this book, I interpret the results of my research in relation to the literature on Third World peasantries. For some people, the category of "peasant" may seem inapplicable to villagers in the Highlands of Papua New Guinea. The popular image of these villagers is one of "tribesmen" involved largely in competitive ceremonial exchanges, tribal warfare, and other "traditional" activities. Similarly, according to a common stereotype, a peasant is a conservative rural cultivator who is severely oppressed by exploitative policies of masters and overlords. As Strathern (1982b) argues, neither characterization alone is accurate in describing the Highlands situation.

However, the stereotype of a peasant noted above does not accurately reflect the current, more encompassing usage of the category in the social sciences. One of the key defining features of peasantries is their relationship with the market, other social classes, and the state. Peasants produce commodities for the market, which enables other classes and the state to extract surplus from them by such means as taxes and unequal terms of trade. However, as researchers in peasant studies have documented, the political and economic conditions under which peasants produce for the market are highly diverse, not uniform, varying not only from region to region but also through time. Thus, some peasants who were once severely exploited by market and state during the colonial era now have considerably more autonomy from these external forces, but researchers still characterize them as peasants. Furthermore, the extent to which peasants are involved intensively in market production often depends on economic boom and bust cycles, which, in turn, affect peasants' economic and political relationships with other classes and the state. Thus, people described as peasants in the development literature face a wide range of conditions; some suffer extreme commercial and political exploitation, whereas others do not. The villagers in this study are not highly oppressed by other classes and the state (though some surplus is still extracted through taxes and unfavorable terms of trade), but their extensive involvement in commodity production makes them peasants. Furthermore, they, like other peasants, rely primarily on household labor, have a low level of technology, and have rights in the land they use. The alternative to using the concept of peasant in my study is to assert that these Highlands villagers are "tribal," but such a characterization ignores the fundamental importance of market and state in the village production process. Although I would not argue that all Highlanders are peasants, many, including the villagers in this case study, clearly are.

I conducted most of the fieldwork on which this book is based from April to October 1976 and January to December 1977 while a doctoral student in the Department of Human Geography at the Australian National University. Although I spent most of that time working in one village, I did travel throughout the Highlands to examine similarities and differences elsewhere in relation to the impact of commodity production. I also returned to the village for shorter visits in August 1979, April 1981, and August 1981. The subsequent trips helped to more clearly highlight the crucial processes at work, particularly the importance of boom and bust cycles.

ACKNOWLEDGMENTS

The Department of Human Geography, Research School of Pacific Studies, Australian National University, generously provided funding for most of the fieldwork. I am particularly indebted to my two supervisors, Diana Howlett and William C. Clarke, who throughout my stay in Australia and in Papua New Guinea offered moral support, unselfish assistance, and intellectual stimulation. I have profited considerably from their advice and criticism. I also want to thank others at the Australian National University. Keith Mitchell produced the maps and figures, and Ken Lockwood provided much assistance in obtaining equipment for fieldwork and in various technical matters. Jim Juvik and Marie Reay made many helpful suggestions concerning data analysis.

I am also indebted to many members of the Commonwealth Scientific and Industrial Research Organization (CSIRO), Canberra. In particular, Pat Walker and Don McIntyre of the Division of Soils and Tjeerd Talsma of the Division of Forest Research provided valuable advice and helped devise field methods for the analysis of the environmental impact of cattle. Data on soil bulk density were analyzed using the computer programs of the Division of Forest Research, Canberra.

Many individuals and government departments in Papua New Guinea were extremely helpful in providing information and assistance. The Chief Archivist of the National Archives and Public Records of Papua New Guinea kindly provided access to past patrol reports. Rick Giddings assisted in the location of historical information in Goroka. At the University of Papua New Guinea, Andrew Wood and Colin Pain of the Department of Geography offered helpful comments on field methodology, and Andrew Wood provided the analysis of soil texture. I am also grateful to many members of the Department of Primary Industry, particularly George Malynicz, Assistant Secretary (Livestock), for his continuing support. Ralph Ernst of the Office of Forests was helpful in locating aerial photographs and in providing enlargements. In Port Moresby, Kainantu, and elsewhere I received much cooperation from many agricultural extension and livestock officers. In particular, Biosi Gunure, Willie Susuke, Aiyuta Aiyako, Allan Ranson, and Allan Marsh aided my research on smallholder cattle projects in the Kainantu area. Toung Duang and his assistants from the Highlands Beef Research Unit at Goroka courageously helped in a census of the Kapanaran cattle herds. Edward Henty and Michael Galore of the Division of Botany identified plant specimens. Members of the Highlands Agricultural Experiment Station, Aiyura, greatly assisted my research effort. Allan Kimber provided access to the facilities of the research station and Stuart Smith identified various insects. My stay in the Highlands was particularly enhanced by the hospitality and friendship of Stuart and Bilha Smith and Bill and Eve van Hecke of Aiyura. Alec Vincent of the Summer Institute of Linguistics provided assistance in translating the Tairora language.

I also want to thank those who have been generous with their time in assisting with the preparation of this book. In particular, Bon Richardson has been a constant source of encouragement and has provided valuable editorial advice. I have also benefited considerably from comments from Diana Howlett, Phyllis Bridgeman, Charles Good, Bernard Nietschmann, Paula Brown, Dan Bradburd, Mike Watts, Mary Beth Pudup, and Terry Hays on earlier chapter drafts. Glenna King and Jane Vyula devoted much time to typing numerous revisions. I acknowledge permission from the *Annals of the Association of American Geographers*, the *American Ethnologist*, and the Papua New Guinea Institute of Applied Social and Economic Research to use previously published material.

Undoubtedly, my greatest debt is to the people of Kapanara whose hospitality and generosity are unmatched.

Peasants, Subsistence Ecology,
and Development in the Highlands
of Papua New Guinea

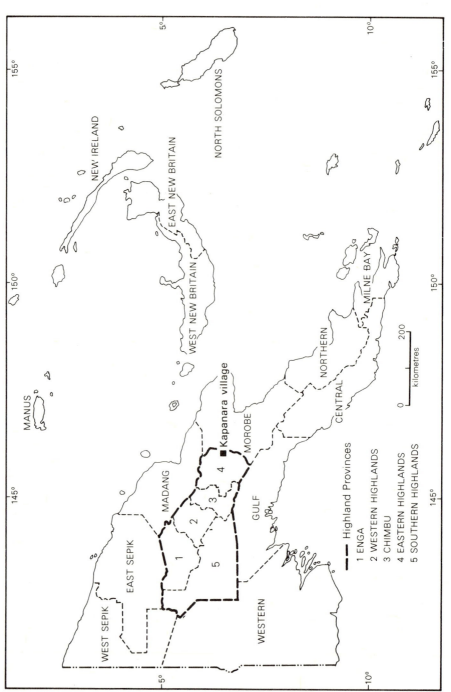

Papua New Guinea

By ANY MEASURE, agricultural commodity production has been and continues to be one of the most pervasive and fundamental forces affecting economic and social life in the Third World. In several regions of the world, participation in cash cropping pre-dates European colonialism, whereas in others it is part of the colonial heritage. In some cases, villagers voluntarily entered into such commercial transactions; at other times, they were forced to do so by colonial government fiat, the introduction of taxation, and the undermining of rural self-sufficiency. Many of today's Third World governments, regardless of ideological orientation, desire to increase rural cash cropping to stimulate their national economy. Countless foreign aid programs financed by such organizations as the World Bank and the Agency for International Development have similar goals.

Planners' and governments' goals are fulfilled when rural market involvement increases. But what is the significance of this change for people in rural communities? When villagers become involved in commercial transactions, they alter their interactions with the local environment, patterns of production, and social relationships. Conflicts between these previously established patterns and commodity production can reduce the viability, welfare, and resilience of rural communities, making them more susceptible to economic and environmental perturbations. Contrary to the assurances of government bureaucrats, planners, and international development agencies, increased market participation does not inevitably lead to improvements in the quality of rural life.

This book is a study of the impact of commodity production in the Highlands of Papua New Guinea, one of the last major regions to be incorporated into the global economy. Changes associated with commodity production that have occurred throughout the world for many generations may be seen in microcosm in the Highlands, where such changes have taken place in only a few decades (Howlett 1980: 195). Two recently introduced rural cash-earning activities—smallholder cattle projects and coffee growing—have been particularly influential in village life. The two activities, typical of rural commercial endeavors widespread in the Third World, can be differentiated by the way they are managed in the village and by the degree of government involvement. Smallholder[1] cattle projects, enthusiastically adopted by Highlands villagers in the 1970s, are heavily influenced by external government policies and regulations. Most are government-designed "projects" financed with development bank loans. They often begin as communal endeavors and require large areas of land. In contrast, the influence of government regulations on coffee production is minimal, with villagers determining the extent of their own managerial inputs. Unlike cattle raising, coffee production does not involve the use of external funds. Villagers grow coffee in small, scattered plots owned and worked by members of individual households.

This study is meant to contribute to several major themes that deal with Third World economic development and peasantries. One basic issue concerns the impact of commodity production on both physical and social elements of rural production systems—the natural environment, subsistence activities, and social relationships. Particular attention is paid to the effects on subsistence production, the major form of livelihood for most peasant households, because the maintenance of subsistence production is absolutely essential for the welfare of rural populations. Another major issue concerns the causes of rural economic differentiation, a ubiquitous phenomenon accompanying the spread of commodity production. A consideration of all these changes in the production process is relevant to understanding the slow rate of change in rural areas in particular and the persistence of the peasantry in general.

In examining these issues, I employ a perspective that I call "the cultural ecology of economic development" (Grossman 1979). The general field of cultural ecology is holistic, stressing the detailed systemic relationships among the numerous variables—environmental, technical, social, and cultural—that affect patterns of resource use and production. It focuses on patterns of resource use and production because they are the primary links between people and their environments; the ability to cope with environmental change is a major determinant of the viability of rural communities. As rural people adapt to the new environ-

---

[1] The term "smallholder" here refers to cattle projects run by villagers in their traditional community setting. The term has sometimes been used by others to refer to people operating a commercial activity on land leased or purchased from the government and outside the traditional community setting.

ments created in the "development" process, they alter both the technical and social arrangements involved in their patterns of resource use and production—changes that have far-reaching significance for their welfare. Using a cultural-ecological perspective to investigate these changes, I analyze alterations among the following interrelated elements: the intricate, complex, reciprocal relationships between people and components of their natural environment; the spatial patterns of local productive activities; the social relations of production; and the influences linking the village community to the outside world that affect resource use and production.

Peasants may be said to adapt to two changing environments—one local, the other external. The physical and biotic components in the village setting are part of the natural environment within the local ecological system. The introduction of plants and animals associated with commodity production, such as coffee and cattle, changes this local environment and can disrupt previously established linkages in the ecological system. The external, political-economic environment also influences the functioning of rural communities, and consists of such varied influences as those from government policies, taxes, moneylenders, commodity prices, credit institutions, landlords, and extra-village sources of wage labor. As the development process progresses, forces emanating from the external, political-economic environment become increasingly important in influencing rural communities. As villages lose their autonomy, many factors affecting human-environment relationships become governed by distant events over which rural communities have little control; these external events usually are insensitive to the needs of rural communities, thus increasing the chance of perturbation (Rappaport 1977; Nietschmann 1979). The variability of this external environment over time has great significance for patterns of change in rural areas.

A cultural-ecological perspective, with its emphasis on detailed, holistic, microscale research, can fruitfully supplement the burgeoning literature on economic development in at least two ways. First, most studies lack an ecological perspective, focusing instead on economic and political issues such as class conflict, the role of the state, capital accumulation, surplus extraction, and opportunity costs. Second, they are usually macroscopic in scale, examining issues at regional, national, and international scales. However, as Samoff (1980: 8) has declared:

results at that level of analysis, however exciting their theoretical innovation, are ultimately frustrating. Too often, the necessary data base is nonexistent. . . . What seems intuitively correct remains empirically elusive.

For example, researchers report declines in subsistence production, but they provide scant data to document the interrelated processes causing the decline at the local level. Such crucial local processes are examined in this study.

SUBSISTENCE AND COMMODITY PRODUCTION

Both development planners and Third World governments—capitalist and socialist alike—stress the need to expand significantly rural commodity production, arguing that substantial national and local benefits would result (e.g. Williams 1978). At the national level, the balance of payments would be aided because export income rises as more cash crops are sold overseas; at the same time, food imports should decrease as rural producers supply more food to urban markets. The nation's industries would also obtain a larger home market for their consumer goods as rural incomes grow. At the local level, improved rural incomes should enable villagers to add variety to the traditional diet, purchase consumer goods to increase their standard of living, invest in agricultural capital, and, where available, employ underutilized land and labor. Where periodic shortfalls in subsistence production occur due to unfavorable weather, cash saved from commodity production would support otherwise hungry households. In addition, growing rural incomes would generate opportunities for village entrepreneurs involved in retailing and other service activities.

Advocates of increased commodity production also assert that if surplus land and labor exist in rural communities, the initial adoption of cash-earning enterprises would not be detrimental to subsistence systems (e.g. Myint 1958, 1969; Fisk 1964, 1971, 1975, 1978; Hogendorn 1975; Meier 1975) and would therefore not cause a decline in traditional living standards. Optimistic planners argue instead that market involvement could only be beneficial for villagers. Because enough land is available to grow both subsistence and cash crops, the area under subsistence crops would not have to decrease. Similarly, time traditionally allocated to local food production would not decline because people could reduce their considerable leisure time, not the time normally employed in subsistence activities, to produce the new commodities. Planners recognize that commodity production would conflict with subsistence endeavors either where high population densities cause land shortages or where cash earning becomes the dominant activity. However, in the initial stages of market participation in areas with surplus land and labor, the impact should be minimal. Sub-Sahara Africa and the Pacific have often been characterized as regions having a surplus in these factors of production, and thus the experience of a village in the Highlands of Papua New Guinea with commodity production is relevant to an evaluation of this perspective.

To accomplish their goal of expanding cash cropping, governments use various measures. They send out agricultural extension agents who attempt to persuade farmers both to increase production and to adopt new cash crops and cultivation techniques. They also provide credit to rural producers for the purchase of synthetic fertilizers, pesticides, and farm machinery, believing that peasants' traditional simple technological inputs are not sufficient to raise production markedly. Governments also raise taxes to spur commodity production, and some even have used force. They justify all these measures for the "good of the nation."

Undoubtedly, segments of the population do benefit from greater commitments to village commodity production. It is becoming increasingly clear that urban groups, bureaucrats, and businessmen are the major beneficiaries (Lipton 1977; Lappé and Collins 1977; Bates 1981). With growth in a nation's export earnings, urban elites are able to import more luxuries to maintain high living standards. More locally produced food entering urban markets reduces food prices, enabling businessmen to pay their urban workers lower wages. Merchants' and crop processors' profits are often directly related to the volume of commodities they purchase from villagers. Government marketing boards, which both set commodity prices and purchase village cash crops in many Third World nations, often keep producer prices artificially low (Bates 1981). They use income from reselling the commodities in national and international markets to support government projects, which largely benefit urban groups, industrial firms, and large-scale capitalist farming interests. Many governments also obtain substantial revenues from taxing commodity exports (Lele 1981). The irony is that rural producers—who are the majority in most Third World nations and should therefore benefit from endeavors designed for the "good of the nation"—not only fail to share adequately in the advantages, but they often are adversely affected by market involvement.

Although the effects of commodity production on rural communities are not uniform throughout the Third World, certain general trends associated with market participation can be recognized. The impact on the natural environment, subsistence, and social relationships—all interrelated elements of rural production systems—can be dramatic and deleterious. The effects on the natural environment and social relationships are well documented, but village-level research on the causes and consequences of conflicts between subsistence and cash cropping is relatively scarce.

Environmental deterioration can result from numerous causes. To increase market output, villagers may shorten fallow periods, cultivate more marginal land susceptible to erosion (Bernstein 1979), or overexploit commercially valuable plants and animals (Nietschmann 1979). In addition, cash cropping sometimes necessitates adopting new cultivation methods such as monocropping, mechanization, or the use of pesticides that may be ill-suited and detrimental to the local environment. The resulting environmental problems, in turn, reduce the potential productivity of both subsistence and commodity production.

Commodity production also disrupts subsistence patterns. The timing of labor inputs into the two forms of production can conflict at certain stages in the agricultural cycle because of the need to plant or harvest in a relatively short period. Commercial agricultural activities often are located closer to settlements, while subsistence gardens are pushed out to more distant, marginal lands. Where land is scarce, commercial activities can reduce the area for local food crops. The encroachment of cash cropping on subsistence is potentially detrimental for villagers for several reasons. Once tied to commercial activities rural producers are at the mercy of fluctuating commodity prices on national and international markets. At the same time, the terms of trade often work to their disadvantage, with the rate of inflation in the price of imported goods higher than that for the primary products sold. Thus, to maintain the same standard of living, people must produce and sell more, thereby increasing the rate of environmental exploitation (Nietschmann 1979). Returns to land and labor are often lower in commodity production (Mitchell 1976; Harris 1978), and hence peasants must use more land and labor to support themselves where cash cropping reduces subsistence output. Subsistence systems, although not disaster-proof, are usually better adapted to local environmental constraints. Furthermore, even if increasing commodity production leads to higher cash incomes, there is no assurance that nutritional levels will also improve because purchased foods, which are often refined carbohydrates, may not be as nutritious as the subsistence items they replace (Nietschmann 1973; Dewey 1981).

Social relationships based on mutual assistance and reciprocity, which help alleviate a household's occasional labor and food shortages, are also undermined by the spread of the commercial economy. They are as much an integral part of the production process as are subsistence techniques, and both often serve the same purpose—ensuring community sustenance and survival (Scott 1976: 2-3). As cash cropping increases, villagers begin selling food they once shared (Nietschmann 1979), and they form communal work groups less frequently (Gudeman 1978). Communal forms of land tenure, which once guaranteed community members access to land, come under increasing pressure as those extensively involved in cash cropping seek more individual forms of tenure to safeguard their investments. Extended family units, which traditionally helped alleviate production problems faced by individual nuclear families, disintegrate as impoverished extended household heads can no longer satisfy traditional obligations to other members (Shenton and Lennihan 1981).

Government officials, planners, and researchers often underestimate the potential negative effects of commodity production on rural communities in general and on subsistence production in particular. They do not adequately consider the numerous variables influencing the relationship between subsistence and commodity production because they focus only on the isolated factors of production of total land and labor available. Indeed, in the present case study, there is a surplus of land and labor; nevertheless, considerable conflict between subsistence and commodity production exists. Using the more holistic cultural-ecological perspective, I argue that a much broader range of local-level variables in addition to total land and labor must be examined: the nature of the linkages within the subsistence system and the local ecological system; the relative location of subsistence and cash-earning enterprises; the seasonality in the demands for labor; rural expenditure patterns; and a community's sociocultural characteristics. Because these fundamental factors are systemically in-

societies, people retained rights in land only as long as they cultivated it. Their subsistence system was based on shifting cultivation in which they used plots temporarily, but when cash crops
were introduced, they began growing perennial tree crops, such
as coffee or cocoa. By continually using the land for commodity
production, they established more permanent land rights. In this
situation, those able to mobilize labor to help plant large areas of
cash crops had a distinct advantage. In many communities, leaders
traditionally obtained labor from followers by persuasion or through
customary obligation, and leaders, in turn, reciprocated by providing protection and redistributing food and wealth on certain
occasions. When commodity production was introduced, some
leaders used their traditional relationship with followers for their
own commercial ends. For example, some clan leaders in the
Highlands of Papua New Guinea persuaded their followers to
plant large coffee estates for them (Finney 1973: 63), and some
traditional African chiefs, taking advantage of labor obligations
owed them by the inhabitants of their district, created large commercial farms (Stavenhagen 1975: 149). Leaders manipulate traditional social obligations to mobilize resources, but when substantial accumulation occurs, they sometimes refuse to honor
their traditional reciprocal obligations to followers (Connell 1979;
Good and Donaldson 1980), thus increasing differentiation.

These three forces all have facilitated rural economic differentiation in the Highlands of Papua New Guinea, but their relative significance in particular areas varies considerably. This case
study emphasizes the implications of government action but departs from most analyses of the role of the state because they
portray villagers only as inert, passive objects affected by official
policy. Thus, I focus not only on the central importance of government activity, but also on how villagers' interpretations and
manipulations of their linkages with the state affect control over
resources and wealth. Concentration on the linkages between
rural communities and the outside world that affect resource use
and production is an essential element in cultural-ecological analyses of development.

Persistence of the Peasantry

The Highlanders that are the center of attention in this book are
peasants (Meggitt 1971; Howlett 1973; Connell 1979; Gerritsen
1979; Fitzpatrick 1980), sharing characteristics with other peas-

antries throughout the world. In such communities, the autonomous household is the basic unit of production, though labor from other households also is obtained occasionally either through mutual assistance or hire. The technology is simple, with reliance on varying combinations of such nonmechanized inputs as the digging stick, hoe, spade, axe, draught animal, and plow. Peasants have rights in the land they use, either as members of traditional communities or as tenants or sharecroppers. People engage in both subsistence and commodity production and often supplement their income with wage labor. Part of a peasant's output is allocated to people outside the household. Some is destined for dominant social classes and the state, which extract peasant surplus through land rent, taxes, *corvée*, interest on debts, terms of trade, and surplus value from wage labor (Wolf 1966; Shanin 1971; Deere and de Janvry 1979). A portion is also used to satisfy community social obligations such as contributions to group ceremonies. The major portion of production—whether in the form of subsistence crops or money from the sale of commodities or wage labor—is usually destined to satisfy a household's basic consumption needs in food, shelter, clothing, and tools. The immediate concern for most peasants is thus the "reproduction" of the household unit itself.[2]

The future of such peasantries has been a major concern for social scientists. Various evolutionary theories, including both modernization theory and classical Marxism, assert that the peasantry is a transitional stage (Williams 1978). With the spread of commodity production, economic differentiation, more productive technologies, and eventually of capitalist relations of production, peasantries are supposed to disappear. Accordingly, some peasants should become full-fledged market-oriented farmers and merchants while the majority become landless or nearly landless laborers in rural and urban areas. Such changes in productive

---

[2] Peasants can be distinguished from capitalist farmers and proletarians. Unlike peasants, who are concerned primarily with household maintenance and survival, a capitalist farmer produces to accumulate capital, reinvest profits, constantly expand his enterprise, and employ more productive technology (Bernstein 1979). Proletarians have lost their rights to use land, and have only their labor to sell to survive. For peasants, the major emphasis on household reproduction is not a function of a behavioral trait eschewing profits but a result of surplus extraction by others that limits their potential for accumulation (de Janvry 1981: 102-106). That economic differentiation does occur within rural communities indicates that peasants can be concerned with accumulation under the appropriate circumstances.

patterns were widespread in much of early industrial Europe. However, although they also occur in the Third World, they are not yet so pervasive as to cause the disappearance of the peasantry. As Williams (1978: 929) notes, "The unwillingness of the peasantry to complete their historical demise has been a fundamental feature of world history. . . ." Similarly, Howlett (1973), in a pessimistic assessment of the potential for socioeconomic processes to radically transform and improve the life of the Highlanders, has coined the phrase "terminal peasantry."

A focus on the forces affecting peasant production patterns in general and commodity production in particular is fundamental to the analysis of why peasantries persist. Because increased market participation would lead ultimately to major transformations in peasant production patterns, the limited extent of rural commodity production is a key variable contributing to the persistence of the peasantry in the Third World. Any explanation of such production patterns must take into account the considerable variability in rural people's commercial involvement. In some places they are hesitant to adopt or expand cash cropping, whereas in others they are keen to do so. In addition, village inputs into commodity production often vary through time, exhibiting "boom and bust" periods. Three interrelated variables influence these diverse patterns: (1) the degree of local autonomy from other classes, the state, and the market; (2) the risks involved in expanding commodity production; and (3) the nature of the external, political-economic environment. Once we understand the significance of these factors, we can then portray how they interact under different conditions to constrain rural change and how they affect the peasantry in the Highlands of Papua New Guinea.

Variations in commodity production are related in part to the degree of local autonomy from other social classes (such as landlords and moneylenders), the state, and the market. Those peasants with greater autonomy can decide whether to increase or decrease production according to their initiative. In contrast, as local autonomy declines, peasants become subjected to the control of external forces that largely dictate their degree of market involvement.

Loss of local autonomy results from several causes. Government policies imposing oppressive taxes or requiring mandatory production of cash crops bind villagers to the market, and extensive indebtedness to usurious moneylenders necessitates continual cash earning to repay debts. The nature of the commodity itself influences autonomy; where land is scarce and peasants have planted inedible perennial cash crops—such as coffee, cocoa, or sisal—on land previously used for subsistence, they must either produce and sell their commodities regardless of price or seek wage labor to compensate for decreased subsistence production. In addition, autonomy declines as people become dependent on the outside world for both technology and consumer goods. Such dependence, which also ties villagers to cash cropping, results from numerous local and external forces: local craft industries are often destroyed by competition from cheaper foreign imports; cash cropping can undermine subsistence production, thus necessitating purchasing food; many "development" projects emphasizing commodity production require the purchase of new, expensive, energy-intensive inputs such as fertilizers and pesticides; at the local level, villagers themselves often want to maintain a certain standard of living based on the consumption of imported goods; and their cash needs increase as money and imported items become essential components of continually inflated ceremonial exchanges, such as bridewealth payments.

Bernstein's (1979) analysis of the plight of African peasants demonstrates the fundamental role of declining local autonomy in relation to exploitative, political-economic forces. He notes that colonial governments did not want to rely on the whims of peasants for needed commodities, so they stimulated continued cash cropping through the imposition of taxation and regulations forcing the planting and production of commodities and by undermining traditional self-sufficiency (Bernstein 1979: 427). As self-sufficiency was destroyed, rural households had to rely more and more on the commercial economy for their reproduction. The "simple reproduction 'squeeze' " often resulted (*ibid.*): prices villagers received from the sale of their cash crops were highly variable, though usually low, but in many cases cash needs increased, partly because of declining terms of trade for the rural populace. Taxes were fixed and sometimes even increased during economic downturns. Thus, when commodity prices fell, villagers either had to intensify the production of cash crops, decrease consumption, or both. Declining returns to labor also resulted as they intensified their agriculture or cultivated more marginal land. Some villagers came to rely on usurious moneylenders to survive periodic shortfalls in either income or production. For some, a cycle of increasing indebtedness ensued, which could result in peasants losing control of their land (see also Dejean 1980). Another peasant response to the squeeze was to shift to less labor-demanding but also less nutritious subsistence crops, such as cassava. In essence, because peasants had lost their autonomy, they were "sucked in" to a commercial system that provided few benefits and many negative consequences.

Loss of autonomy is not inevitably associated with continued cash cropping. Elements of the external, political-economic environment sometimes suppress village cash-earning endeavors to benefit capitalist interests. For example, in the early colonial era in Kenya and Zimbabwe, villagers responded on their own initiative to the growing European-settler market by rapidly expanding cash cropping. However, colonial administrators subsequently restricted peasant commodity production to limit competition with expatriate planters, thus forcing Africans to earn money by providing cheap rural labor for European farmers (Palmer and Parsons 1977; Porter 1979).

In contrast, a high degree of local autonomy gives peasants greater latitude in making decisions regarding their inputs into commercial activities. Thus, when commodity prices drop, they can more readily decrease cash cropping than can those whose autonomy has been reduced by external forces. Several conditions contribute to peasant autonomy. Weak national governments often are unable to enforce policies designed to reduce peasant autonomy. In some areas, as in Papua New Guinea, rural taxes are relatively low and usurious moneylenders are unimportant. Peasants who cultivate their own land, unlike tenants and sharecroppers, are not obligated to produce commodities to satisfy landlords. Where subsistence production is both plentiful and reliable, the need to purchase food is much less. People selling edible annual crops are less affected by declines in commodity prices because they can consume their product and plant different crops the next year. The maintenance of a viable social network based on reciprocity and mutual assistance providing security in times of need also increases rural autonomy. Under such circumstances of limited dependence, people are able to reduce or expand their involvement in commodity production according to their own initiative.

In deciding whether to increase cash cropping, peasants must consider the degree of risk involved in allocating resources to commercial endeavors. Villagers may limit commercial activity to minimize the risks posed by their environments (Wharton 1971). Ecological problems such as periodic drought or flooding can make agricultural output highly variable and often very marginal, with occasional serious food shortages. These risks have become exacerbated where colonialism and the commercial economy have undermined traditional agronomic techniques (for example, polyculture and planting crop varieties with different drought resistance) and social means (such as food sharing and communal work groups) of coping with natural hazards and where dominant classes have expropriated much of the peasants' surplus (Scott 1976). Where peasants face these problems and either land or labor are scarce resources, adopting locally untested commodities at the expense of subsistence crops can be very risky because people are uncertain whether the innovations will provide returns comparable to well-adapted subsistence crops whose yields are more predictable. Peasants are unsure about the new commodity's susceptibility to drought and disease, its most appropriate cultivation techniques in the local environment, yields, market price at harvest time, and future cost of required inputs (Wharton 1971; Guillet 1981). If the cash crops do not produce the expected returns, the likelihood of severe deprivation or even starvation is greater. Thus, where individuals live close to the subsistence margin, they are likely to minimize risks rather than gamble to maximize potential, but uncertain, cash profits, giving rise to what Scott (1976: 15) calls the "safety-first" principle—peasants are concerned first and foremost with providing adequate subsistence, and only then will they engage in more risky market transactions (see also Guillet 1981: 6). Although risks are not an absolute barrier to increased market involvement, particularly because peasants can reduce potential problems by pooling land, labor, and cash (ibid.: 9), they certainly can inhibit change in rural production patterns.

Peasants face additional obstacles in relation to numerous elements in the external, political-economic environment. Those who are tenants may lack security of tenure, and thus often hesitate to make investments required to expand commodity production. Commodity prices are highly variable, but often very low, thus providing peasants with little incentive to produce cash crops. Villagers may hesitate to accept advice from agricultural extension officers extolling the benefits of new cash crops because past government extension policies have been perceived as either detrimental to their welfare or simply unrealistic (Williams 1976: 132-133). Where extension officers are perceived more positively, a poorly funded, weak agricultural extension service, as is common in many parts of Africa (Lele 1981), hinders cash-cropping expansion. In addition, many governments fail to adopt policies that would facilitate more commodity production, such as making land a legally marketable commodity or providing adequate credit to rural producers; because land cannot legally be bought and sold, rich peasants who have accumulated capital may find it difficult to reinvest in and expand agricultural production, and they therefore invest in nonfarm endeavors such as transportation, crop trading, retailing, and education (Bernstein 1981: 52). The influence of all elements of the external environment is rarely uniform, however, as they often conflict with each other. For example, a government may design a program for commodity production and make an intensive extension effort to increase cash cropping, but low crop prices may thwart its efforts.

In contrast to the perspective I have outlined, adherents of the conventional, modernization approach have asserted that peasant values and cultural characteristics are the main hindrances to commercial activity (e.g. Foster 1973). Peasants supposedly do not expand commodity production because they are inherently conservative, bound by tradition, and do not take chances. Limited aspirations, religious beliefs, fear of envy, the threat of sorcery, or other social sanctions inhibit individuals from accumulating capital. This perspective fails to explain the widespread evidence of a positive peasant response to increase cash cropping when producer prices rise, especially when accompanied by appropriate government policies (see Hutton and Cohen 1975; Bates and Lofchie 1980; Bates 1981). By focusing on internal village-level constraints only, the approach ignores the more crucial external, political-economic forces hindering cash cropping. Certainly, traditional cultural values and social patterns affect commercial activity, but they are not the predominant influence; indeed, sometimes the values and attitudes that constrain market involvement—such as mistrust of outsiders—are not "traditional" but have evolved in response to harmful interactions with the external political-economic environment (Hutton and Cohen 1975). In addition, traditional values may actually facilitate commodity production. For example, in the Highlands of Papua New Guinea, the traditional, precontact cultural emphasis on wealth accumulation, status achievement, and competition has spurred rural cash cropping (Finney 1973).

The degree of local autonomy, the risks posed by expanding market endeavors, and the nature of the political-economic environment are thus the essential variables influencing the extent of rural commodity production.[3] To explain the considerable variation in peasant responses to the commercial economy, we must now indicate how these variables interact. To apply this perspective to particular case studies, situations having different degrees of autonomy must be considered separately.

Peasants with limited autonomy often are bound to other classes and the state through exploitative relationships. Although commonly forced to produce a certain amount of cash crops, such peasants usually hesitate to expand cash cropping beyond a certain level because of local environmental risks, which are often magnified because traditional means of coping with natural hazards have been undermined. In addition, the external, political-economic environment is not conducive to increased commercial activity. Peasants, for example, may realize that their cash-cropping efforts often will benefit other classes and the state more than themselves. However, when crop prices fall and the simple reproduction squeeze occurs, peasants may be forced either to produce more commodities or to engage in more wage labor to survive. At the same time, economic differentiation often increases as poor peasants lose control over some of their productive resources.

Where a considerable degree of autonomy exists, local environmental risks can also hinder commercial activity, but when such risks are comparatively low, other factors affect change in productive patterns. This case study of a village in the Highlands of Papua New Guinea helps illuminate the processes of change in an area of low environmental risk and a high degree of local autonomy. In such cases, the rate of change in productive activities is related ultimately to the highly variable nature of the linkages between rural communities and their external, political-economic environment. Highlands villagers respond positively to both commodity price increases that they believe are rewarding and to extensive government extension inputs, which in the past have generally been perceived as beneficial. When these two external factors provide incentive, peasants shift toward more cash cropping. This shift, in turn, brings about certain perturbations at the village level that often function as positive feedback cycles

[3] Undoubtedly both a low level of technology and limited land availability can restrict commercial output, but it is the lack of financial returns from commodity production—a function of the external, political-economic environment—that ultimately frustrates attempts to surmount these obstacles.

to reduce subsistence production and to bind people even more forcefully to the commercial economy. Economic differentiation also increases. However, when these external influences no longer provide the stimulus because of declines in commodity prices and extension inputs, commercial activity and associated perturbations decrease because peasants, having a high degree of autonomy, can reverse their commitment to the cash sector. If external influences continue to weaken, cash cropping eventually will reach a low point defined by a minimally acceptable rate of consumption of store-bought goods, by basic household cash requirements (such as taxes and school fees), and by social obligations to maintain status and prestige. However, some perturbations persist, creating continuing stresses on subsistence production. In addition, although the rate of economic differentiation also slows, it rarely returns to pre-existing levels. Nevertheless, precisely because the intensity of inputs from external sources is so highly variable through time—producing characteristic "boom and bust" cycles—the potential for Highlands peasantries to undergo major structural transformations is thereby limited.

The Incorporation of the Highlands

The incorporation of the Highlands into the global economy is very recent. In the late 1800s, the British and Germans were colonizing the southern and northern coastal areas respectively of what is now Papua New Guinea. They were unaware that several hundred thousand people, living mostly at altitudes between 1,500 and 2,100 meters, inhabited the intermontane valleys and plateaus in the rugged, mountainous chain running from east to west through the center of the island. By the time the Australian government obtained possession of both the British and German colonies in 1914, Europeans[4] still had not entered the region. Missionaries first visited the eastern edge of the Highlands in 1919 (Radford 1972), but major exploration of the area did not occur until the 1930s when Australian government patrol officers, missionaries, and gold miners penetrated much farther to the west.

The Europeans found numerous, small-scale, tribal societies.

[4] The term "European" is used to refer to all those in Papua New Guinea who have European ancestry, such as the Australians, British, and Germans. Most of the "Europeans" who explored and settled in the Highlands were Australian.

Highlanders' subsistence was based on horticulture and pig hus-bandry, supplemented by hunting and gathering. In the Eastern Highlands, their well-tended, rectangular gardens prepared in un-dulating grasslands greatly impressed the Europeans, particularly because they were cultivated without the aid of steel tools. Throughout the region, settlements consisted of, at most, several hundred people and were both aggregated and dispersed. Villagers lived in circular or oval houses with walls made of wooden planks, bark, and grass, and covered with a grass roof. Intervillage warfare was chronic. Land shortages contributed to tribal fighting in areas of high population density in what are now the Simbu[5] and Enga Provinces. In most of the Highlands where densities were much lower and land abundant, fighting settled quarrels over theft, women, and pigs, and avenged the spirits of those killed in battle or by sorcery. To protect themselves from the ever-present danger of attack from nearby enemy communities, their settlements were enclosed by tall, wooden palisades. Adult men lived in a large house where they planned activities, laid strategy for defense and warfare, and performed secret rituals. Women inhabited smaller, separate huts with children and sometimes their pigs. Settlements were mostly self-sufficient, though people had to rely on inter-group trade to obtain locally unavailable essentials such as salt and stone axes and luxuries such as colorful plumes and shells. Knowledge of the world outside the small community was ex-tremely circumscribed; in most cases, individuals rarely travelled far beyond their own territory or that of nearby allies because of fear of warfare and sorcery attack. Communities were relatively autonomous, and there was no state and no enduring political units or political offices. Community leaders were "big-men" who acquired status by prowess in warfare, sorcery, competitive cer-emonial exchanges, oratory, and dispute settlement. They main-tained influence over their own settlement and sometimes ad-jacent ones only so long as they retained their skills, ability, and level of dynamism. Big-men created military alliances with other communities, but like their own tenure as leaders, alliances were unstable and impermanent.

For the Highlanders, the arrival of Australians and other Eu-ropeans opened up an entirely new world. Brown's (1972: 66) description of its significance for the Simbu is applicable to High-landers in general:

5 Formerly known as Chimbu Province.

This sudden entrance of representatives of 20th-century Australia had no precedent in tribal life. Almost everything about the newcomers was different, their bodies, color, clothing, equipment and supplies, food, language, behavior, relationships with one another and with the local people. They communicated by radio and air with a world beyond the mountains of which the Chimbu had no experience. It was their first indication of the existence of this external, white man's world.

Before World War II, the Australian colonial administration's primary objective was to establish government stations in the region, and from these posts to send out patrols to make contact and pacify the Highlanders, who appeared to be in an anarchic state of intergroup warfare and hostility. The reaction of Highlanders to the early administration patrols was varied. Some acted violently, attacking the European intruders with bow and arrows and stone axes. Many feared the white-skinned Europeans as ghosts of their ancestors. Others were friendly. Whatever the initial reaction, pacification proceeded gradually, accomplished by a remarkably small number of Australian patrol officers and indigenous policemen, whose superior firearms, material wealth, and technology greatly impressed the local population. Although some Highlanders were killed during the early skirmishes, the pacification process did not have the dramatic destructive effect on the Highlands population that it had on colonized peoples in other parts of the world.

Nevertheless, Highlanders experienced important changes. Those near European outposts began supplying labor, food, and firewood to the newly established administration stations and missions, to prospectors who unsuccessfully sought major alluvial gold deposits, and to Australian patrol officers and their indigenous policemen who travelled through the region attempting to establish order. Highlanders moreover helped construct airstrips, the only transport link to the more-developed coastal areas, and expanded many traditional footpaths to bridlepaths. For their efforts, villagers received new trade goods such as steel axes and knives, twist tobacco, cloth, beads, and salt, as well as highly prized shells, a traditional form of payment previously obtainable only through long-distance trading links to the coast. With the gradual cessation of warfare, the introduction of more efficient, time-saving steel tools for use in subsistence activities, and the dramatic increase in the number of shell valuables in the Highlands, people channelled much of their energy to social activities. They increased the scale and frequency of interpersonal and intergroup competitive feasts and ceremonies in which they exchanged garden produce, pigs, nuts, forest game, and traditional valuables such as shells and plumes. Because the massive influx of shells caused devaluation, traditional payments involving them became inflated, resulting in hardships for remote communities that could not obtain shells from Europeans (Hughes 1978).

Missionaries—Lutheran, Catholic, and Seventh-Day Adventist—and their coastal evangelists set out to convert and "civilize" the people, exhorting them to abandon such customs as male initiation and other secret rituals, their belief in sorcery, and polygamy. Mission preaching, to some extent, undermined people's confidence in their own world view, though they retained many of their traditional beliefs. The denigration of traditional beliefs and customs, coupled with the overwhelmingly impressive material wealth and sophisticated technology of the intruders, contributed to a feeling of inferiority in relation to Europeans.

During World War II, Highlanders escaped the major destruction that was inflicted on the north coast and islands after the Japanese invasion in 1942. Allied troops were stationed in the Highlands, but only minor military action occurred there. The major negative impact was a dysentery epidemic, which was brought from the lowlands by soldiers and killed several thousand villagers (Finney 1973: 28). During the war, colonial administration efforts at pacification and control were relaxed, and in many areas intergroup warfare flourished again.

Change accelerated after the war. Administrative control was reestablished and pacification of most remote areas was accomplished by the mid-to-late 1950s. Towns grew around administrative centers and their airstrips. Starting in the late 1940s and early 1950s, villagers received cash as payment for their produce and labor and were thus able to purchase much-desired imported goods, such as steel tools, cloth, matches, and salt, as well as shells, at European-owned trade stores in towns and in mission outposts. They also began using cash, along with traditional valuables, in ceremonial exchanges such as bridewealth payments. The colonial administration imposed taxes in the 1950s, which had to be paid in cash, spurring villagers' involvement in commodity production and wage labor. Both secular and mission schools opened, and Highlands children, both boys and girls, received their first formal schooling.

European settlers were the first to raise export cash crops. They established small coffee plantations in the Eastern and Western Highlands, leasing land that villagers sold to the administration. In the Eastern Highlands near the administrative center of Goroka, for example, thirty European settlers leased properties ranging from 15 to 166 hectares by the mid-1950s (Howlett 1962: 229-230). Highlanders usually alienated uninhabited, fertile valley bottom land once fought over and claimed by several villages. Initially, they were pleased to sell their land so that Europeans could settle among them, looking upon the foreigners as potential purchasers of surplus food crops and labor and as a source of knowledge concerning the attainment of the great wealth held by Europeans (Finney 1973). However, many villagers were subsequently disillusioned with the resulting meager benefits. Land alienation occurred mostly in the early 1950s, but then the administration curtailed land purchases markedly in response to harsh criticism of its policy and fears that village land shortages might result. It was never as extensive as it has been in the lowland areas, and certainly the overwhelming majority of land in the Highlands—99 percent—is still controlled by villagers themselves (PNG, National Statistical Office 1982: 57). By 1974, there were 90 Highlands coffee plantations with a total planted area of only 7,000 hectares (Munnell and Densley n.d.: 29). However, the best agricultural land was often sold, and in a few localities land shortages for village commodity production have resulted (Howlett 1973).[6]

Villagers also became involved in export commodity production. Planners felt that the negative impact of the introduction of such cash cropping would be minimal because of the surplus land and labor in most Highlands areas. Indeed, traditional subsistence systems there have been described as being in a state of "subsistence affluence" (Fisk 1964, 1971, 1975; see also Lam 1982):

a condition in which population pressure on land resources is relatively light, productivity per unit of applied labour (as distinct from available labour) is very high, and most subsistence agriculturalists are able to produce as much as they can consume (with satisfaction) of their main essential requirements,

[6] From 1974 to 1983, original landowners and nationally owned business groups have purchased approximately 65 coffee plantations from European owners in the Highlands using funding from the government and bank loans (Eaton 1981, personal communication 1983; R. M. Mitio, personal communication 1983).

and to sustain an adequate level of living by their traditional standards, at the cost of as little as fifteen or twenty hours labour a week. (Fisk 1971: 368).

Village export cash cropping became significant in the 1950s, when agricultural extension officers made intensive efforts to introduce coffee growing to Highlanders, who quickly adopted it, often planting the crop on prime land. To assure a market for the coffee crop at a time when a world surplus was developing, the Territory[7] along with Australia ratified the International Coffee Agreement of 1962, and one requirement of the Agreement was cessation of extension work promoting coffee planting (Finney 1973: 66).[8] Nevertheless, villagers on their own initiative continued to plant coffee, and the crop spread rapidly throughout the countryside, becoming the major source of village cash income. Highlanders have gone through periods of "hope and disillusionment" in relation to coffee production, initially motivated by unrealistic financial expectations, and subsequently influenced by changes in crop prices, technical processing innovations, and new marketing arrangements (Brookfield 1973: 153). Other introduced export cash crops such as tea, passionfruit, and pyrethrum have met with little success, usually because of poor economic returns or technical difficulties. Highlanders also started to sell traditional and new food crops such as sweet potato, maize, and peanuts at newly established markets in towns and rural areas. Agricultural extension officers also introduced both smallholder cattle and piggery projects in the 1960s to help diversify the rural economy. In the 1970s, the number of village cattle projects grew rapidly with the aid of development bank credit and a vigorous agricultural extension effort.

Highlanders increased their involvement in wage labor as well. Although some worked in plantations or agricultural experiment stations in the Highlands, most post-war employment occurred in the lowlands. Initially, the administration was opposed to sending Highlanders to work in the lowlands, fearing that village social life would be disrupted and people would be exposed to new health risks, but the rising demands of coastal plantation owners for

[7] Papua New Guinea was known as the Territory of Papua and New Guinea until 1971.

[8] Starting in 1973, when international coffee quotas were suspended, the agricultural extension service again began encouraging villagers to plant coffee (Anderson 1977: 4).

cheap, unskilled labor eventually changed its policy (May and Skeldon 1977: 5). The administration thus established the Highland Labour Scheme in 1949, under which it recruited thousands of men for two-year contracts to work on coastal cocoa, copra, and rubber plantations until the scheme ended in 1974. Initially, Highlanders were keen to participate in the scheme, and where recruiting was extensive, the resulting shortage of male labor produced strains in the subsistence system (Boyd 1981). The scheme mainly benefited the expatriate planters, who gained access to a large supply of cheap labor. The coastal working experience provided little useful agricultural knowledge, as lowland plantation crops are not suited to the Highlands climate (Ward *et al.* 1974), though some villagers obtained valuable skills such as truck driving. Highlanders became more familiar with the outside world and the commercial system. They also brought back cash, items such as steel tools, pots, and gifts, and they introduced novelties such as radios, watches, and the now-popular custom of card playing. Villagers' enthusiasm for the scheme gradually waned due to the monotonous and demanding work routine, harsh treatment by some racist plantation managers, low pay, and increasing cash-earning opportunities at home. Recruiters thus had to search continually farther afield to find willing recruits. As participation in the scheme declined, villagers on their own initiative sought jobs in plantations and urban areas, though urban employment opportunities were and still are particularly limited. Another change in employment patterns is now occurring as villagers themselves hire others to help in such activities as coffee production and cattle raising (see Howlett 1980). Though not a predominant feature at present, such employment will grow in the future as rural economic differentiation and land pressures increase with the expanding commercial economy.

The rural population has also been significantly involved in other commercial activities since the 1960s. Some have used coffee income to purchase trucks, transporting fee-paying passengers between village and town. Truck owners also have become involved in the lucrative enterprise of purchasing coffee from villagers and reselling it to processing factories. Many others have established small trade stores in their own community, selling a limited number of goods such as canned foods, rice, matches, soap, and clothes. Such endeavors represent a significant expansion of the commercial economy in rural life.

An extensive road network has greatly stimulated economic

growth. The most significant accomplishment has been the construction of the Highlands Highway, which provides the only road link to the lowlands. The Highway greatly reduces the cost of transportation, because Australians previously had to fly supplies into the Highlands. The initial section was completed in the mid-1950s but was not open to freight traffic until 1966 (Young 1973; 1977). Since then the road has been improved and expanded to reach the western and southern Highlands areas. Now a steady stream of freight trucks moves goods into the Highlands and carries away the processed coffee crop to the lowlands for eventual export. The administration also required Highlanders to construct numerous unpaved roads linking villages with towns and administrative centers. The transport network grew as some villagers built their own feeder roads for better access to coffee buyers and the outside world. The road network is most extensive near the Highlands Highway, administrative centers, and towns. The more remote areas are not yet accessible by road and are linked to the outside world only by irregular air service or by traditional footpaths.

Prospects for development are also influenced by demographic patterns, which have been modified during the Australian colonial era. The cessation of warfare, the breakdown of traditional sex taboos, and the introduction of modern medical services all have contributed to increased population growth, now approximately 2 percent in the Highlands provinces (PNG, National Statistical Office n.d.). Rural densities are as high as 300 per square kilometer (Howlett *et al.* 1976: 28), and in traditionally crowded regions commodity production exacerbates land shortages. The potential for landlessness or inadequate access to sufficient land to satisfy a household's subsistence needs certainly exists in such places (*ibid.*: 116), and will likely eventually contribute to the development of a rural proletariat (Howlett 1980). In most of the Highlands, however, population pressure is not nearly that severe. Densities in rural communities vary greatly, but are usually less than 50 per square kilometer, though even in relatively uncrowded areas, commercial activities—cattle projects, in particular—can contribute to localized land shortages for those wanting to expand commodity production.

These changes have not occurred uniformly throughout the Highlands. Areas close to the early administrative centers were pacified before World War II, while more remote areas were not under control until the late 1950s or early 1960s. Mission influence is strong in some areas, negligible in others. Regions with suitable land and adequate marketing infrastructure rely on cash cropping as the major source of income, whereas areas that are more remote and have limited cash-earning potential at home provide migrants seeking wage labor in plantations and towns. Regional economic differentiation within the Highlands is also increasing, with the Eastern and Western Highlands Provinces having the largest rural incomes due to extensive village coffee production.

Without doubt, the colonial era, which ended in 1975 with independence, was much less disruptive and traumatic in the Highlands than in the coastal areas of the country or in other parts of the Third World. The Australian colonial administration did not wage devastating wars of conquest nor forcibly expropriate land. Neither did it impose regulations requiring the planting and production of cash crops, though the imposition of taxes spurred both cash cropping and wage labor. Other factors have also helped to mitigate the effects of colonialism. The extent of surplus extraction by other classes and the state has been low compared to many other Third World countries (see Lam 1979). For example, head taxes are relatively low, and nonpayment is common where people are disillusioned with their local governments. Usurious cash debt, often an outgrowth of people's inability to meet basic household needs and a major contributing factor to underdevelopment, is absent in the region. The degree of local community autonomy remains high, and the commitment to commodity production is largely reversible. Although drought and frost are occasional localized problems (Waddell 1975), risks posed by the natural environment are not extensive. Certainly, the widespread poverty common in Africa, Asia, and Latin America does not exist in the Highlands.

The relatively benevolent nature of the colonial experience, however, does not mean that the Highlanders will escape the negative consequences associated elsewhere with the expanding commercial economy. In considering conditions in the Highlands, the recency of incorporation into the global economy must be remembered. Nevertheless, certain disturbing trends exist. The Highlands commercial economy is precariously dependent upon a single export cash crop, coffee, which can account for up to 90 percent of rural commodity income (Munnull and Densley n.d.: 7) and the price of which is highly variable over time. Rural economic differentiation is increasing. In areas with extensive commodity production, wealthy villagers are pressuring for more individual forms of land tenure, and unofficial leasing and selling of village land, although not widespread, are increasing (Ward 1981). Dependence on imported processed foods in rural areas is also growing. Processes associated with the expansion of the commercial economy are likely to intensify because the government is firmly committed to expanding rural commodity production, and furthermore, villagers themselves want more opportunities in cash cropping. Such processes, in turn, can significantly reduce the degree of local autonomy. I now turn to one village in the Eastern Highlands that was enthusiastically committed to the commercial economy in the mid-1970s to highlight in more detail the impact of the development process on rural communities.

CHAPTER ONE

Kapanara Village

THE *Bisnis* ETHOS

As IS TRUE for peasants in rural communities throughout the Third World, the fortunes of the people of Kapanara village (see Map 1.1) are linked intimately with the commercial economy. They have been involved in cash-earning enterprises, or *bisnis* in *Tok Pisin*, for approximately 25 years. They have earned money from mining gold; growing passionfruit, vegetables, and coffee; operating trade stores and passenger trucks; and most recently, raising cattle. Their commitment to these activities has varied considerably through time, reflecting oscillating periods of enthusiasm and disillusionment. In terms of the amount of land and labor utilized and revenues produced, cattle projects and coffee production are currently their most important commercial endeavors. Influenced by high coffee prices in 1976 and 1977 and high expectations about the potential returns from cattle raising, the villagers made a substantial commitment to both forms of commodity production. This intensive commitment was not the result of indebtedness to other social classes or oppressive government policies, problems experienced by many peasants in other parts of the Third World. Kapanarans still retain substantial autonomy, and thus to a considerable degree can increase or decrease commodity production according to their own initiative. To explain their involvement we must probe into the *bisnis* ethos that pervades the community.

Cattle raising, coffee production, and other commercial enterprises are not only sources of income. They are also symbols of *bisnis*, presently a culturally valued endeavor. *Bisnis* has entered into the criteria defining a man of respect and prestige, one held in esteem by his fellow villagers. To earn respect a man must not only help his kin, be generous, work hard, and provide adequately for his family. He must also be a man of *bisnis* who is engaged in various cash-earning enterprises. What is significant for the man of *bisnis* is not so much the amount of money he makes, though this may be important, but that he is involved in *bisnis*. Such activity, in and of itself, is considered good. This is regarded as a self-evident truth. It is evidence of one's character. Not to

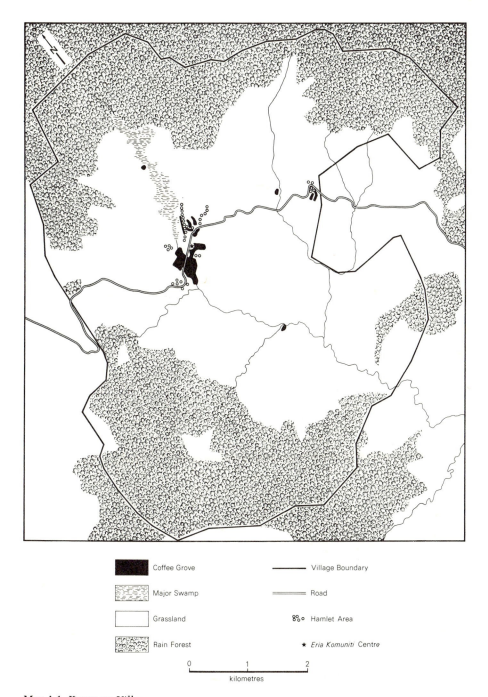

▓ Coffee Grove		———	Village Boundary
Major Swamp		═══	Road
Grassland		8°₀	Hamlet Area
Rain Forest		★	*Eria Komuniti* Centre

0   1   2

kilometres

Map 1.1. Kapanara Village

be involved in any form of cash-earning activity invites the label of *baehi baiinti* in Tairora or *rabisman* in *Tok Pisin*, a person of no worth or consequence, lacking in prestige.

I am not arguing that people will engage in any *bisnis* activity no matter how unprofitable. The villagers abandoned passionfruit growing and goldmining because either the returns were poor or they considered the labor inputs too demanding. However, periodic disillusionment in the past does not negate the belief today that *bisnis per se* is a worthwhile activity. No one questions whether *bisnis* is a good thing. Rather, the question is which cash-earning enterprises are good.

The villagers' motivations for participating in commodity production and other commercial activities are complex. An understanding of these motivations is necessary to explain the perpetuation of the *bisnis* ethos.

The most frequent answer to questions concerning reasons for involvement in the cash economy is that people want to earn money. They need cash to pay taxes and school fees, for transportation, and to purchase certain necessities and items of consumption that have become part of the currently accepted standard of living. Cash is also necessary for full participation in social life, and village life is intensely social. Money is required for such varied activities as beer drinking, gambling, contributing to bridewealth and compensation payments, and participating in competitive exchanges.

People do not accumulate money for hoarding or purchasing substantial amounts of material goods for themselves. They channel much of their income, either as cash or goods purchased with money, into the system of reciprocal exchange, in which generosity is highly valued. The more money an individual has, the greater is his potential to give to others. Contributing to another's bridewealth payment, helping a relative with a feast for his affines, giving generously in exchanges, and providing plentiful food to guests are manifestations of such valued behavior. In addition, giving to others creates an obligation for reciprocation. If an individual has several others in his debt, he is able to call upon them for help in certain endeavors designed to increase his prestige.

Kapanarans do not channel all of their cash into the exchange system. They also reinvest in enterprises such as cattle projects and trade stores to increase their income. Although reinvestment enlarges an individual's potential to give to others, it also creates the possibility of greater capital accumulation and hence greater economic differentiation. The continuing satisfaction gained from participating in the reciprocal exchange system and the strong emphasis on sharing, which is backed by the fear of social sanctions such as sorcery, somewhat discourage extensive capital accumulation. Villagers often attribute sickness in humans and their pigs to sorcery carried out by someone who is angered by the failure of the victim or the victim's owner to share adequately. Similarly, elsewhere in the Eastern Highlands, many Gorokans attributed the deaths of several prominent national[1] businessmen to sorcery performed by persons envious of the commercial success and high status of the deceased (Finney 1973: 113). Nevertheless, even with these constraints greater differences in wealth are emerging among Kapanarans.

The popularity of *bisnis* is strengthened by a traditional characteristic of the Highlands population, interpersonal and intergroup competition. Success in competition brings renown and prestige, and competition in *bisnis* activity is one arena in which people can gain prestige. Groups that adopt new enterprises such as cattle projects gain prestige by favorable comparison with those groups that have not done so. This competitive spirit was demonstrated clearly in Kapanara when one group decided to pool money to purchase a passenger truck[2] after learning that another group in the village was going to buy one. A villager stated his reason for pooling the money: "It is not good that we see another kin group start a business, and they stand up as men and we just feel shame." The size of the enterprise is also an important consideration in competition. The group with the most cattle in its project, for example, gains prestige in relation to those that have fewer animals. In essence, investments in commercial endeavors such as cattle raising, coffee production, and trade stores provide clearly visible evidence to impress others, demonstrating to them one's involvement in the valued and prestigious realm of *bisnis*. Public displays of wealth have traditionally been an integral part of interpersonal and intergroup competition, and in the case of commercial activities, group pride and prestige are enhanced by what Finney (1973: 80-81) calls "conspicuous investments." The phrase "There's no business like show business" may be appli-

---

[1] The word "national" refers to any person whose ancestors were Papua New Guineans.

[2] Trucks that are licensed to carry passengers for a fare are commonly known as PMVs or public motor vehicles.

cable to Broadway, but the sentence "There is no show business like *bisnis*" is clearly applicable to the Highlands of Papua New Guinea.

Stimuli for the adoption of commercial activities also lie outside the traditional system of relationships and exchange. Economic development is official government policy, and occasionally government officials from Kainantu,[3] the District headquarters, visit Kapanara to stress the importance of business, self-reliance, and economic development for the good of their newly independent nation. As inspiration, they present accounts of the economic progress of villagers in other Third World countries. Other government agencies, most notably the Department of Primary Industry, have been instrumental in influencing people to start such enterprises as coffee growing and cattle raising. In addition, the elected Councillor representing Kapanara often returns from local government council meetings in Kainantu with reports concerning new programs in economic development, how business will help the country, and the part that Kapanara must play. Whatever the source, the message is the same: *bisnis* is *good!*

Commercial enterprises were originally a European activity. The villagers' understanding of and relationship with Europeans provide another clue to their participation in income-earning endeavors. Highlanders were often awed by and envious of the superior technology and wealth of the Europeans. To emulate Europeans in activities such as business enterprises demonstrates to fellow villagers one's prowess, ability, and competence in accomplishing feats traditionally performed only by Europeans. (The speech of one cattle project leader in Kapanara reflected this motivation. While intoxicated and distributing beef at a party, he boasted to those assembled that he was just like the white man because he had many cash-earning enterprises, including cattle, coffee, a trade store, and formerly a passenger truck.) In addition, the racist, paternalistic, and intolerant treatment of villagers by some Europeans left no doubt in the people's own minds that they were perceived as being inferior by at least a segment of the white population. To adopt income-earning enterprises was one way of achieving equivalence with the Europeans in the people's own view. Furthermore, to the Kapanarans, the European's life appears relatively luxurious and easy when compared with the small amount of physical labor he performs. In contrast, the vil-

---

[3] The population of Kainantu in 1980 was 3,779.

lagers view their own lives as just the opposite, consisting of hard work, which is sometimes physically painful, yielding little in luxury or comfort. Cash-earning activities, the European way, are thus perceived as being a more satisfactory way of obtaining a livelihood.

Although *bisnis* is a fundamental aspect of the people's lives, it is not their major form of livelihood. Kapanarans traditionally were and still are primarily horticulturalists, growing most of the food they consume. They exploit several culturally recognized environmental zones and plant a variety of crop combinations particularly suited to the different areas. Grown in small, convex mounds, the sweet potato is not only the main staple but also the favorite food. Gardens are fallowed for different periods depending on the type of crops planted and the environmental zone.

Food production is not solely for human consumption. People feed a considerable amount of the garden produce, mostly sweet potato, to their domesticated pigs. Without fail, the daily round of events is punctuated by the squealing pigs hungrily awaiting their morning and late afternoon rations provided by the women of the household. Most families own a few pigs, with the more enterprising and industrious households caring for ten or more. Owning many pigs provides prestige. Villagers also value their pigs because pig meat and fat are prized delicacies and are used as remedies for illness and pain. In addition, pigs are an integral part of the gifts presented in the exchanges and ceremonies recurrent in village life.

Given the dedicated commitment to commercial enterprises and the need to engage in subsistence endeavors, a possible conflict between the two forms of production arises. This conflict certainly is not unique to Kapanara, but is widespread in rural communities throughout the Third World. However, before considering this issue in more detail, we need a more in-depth view of the local setting.

THE VILLAGE SITE

The village of Kapanara is in the Tairora Census Division of the Kainantu District, Eastern Highlands Province (see Map 1.2). Situated on the easternmost edge of the Central Highlands, the villagers are speakers of the Tairora language, which is part of the Eastern Family of the East New Guinea Highlands Stock (Wurm 1978), the largest language stock in the Highlands. The boundary

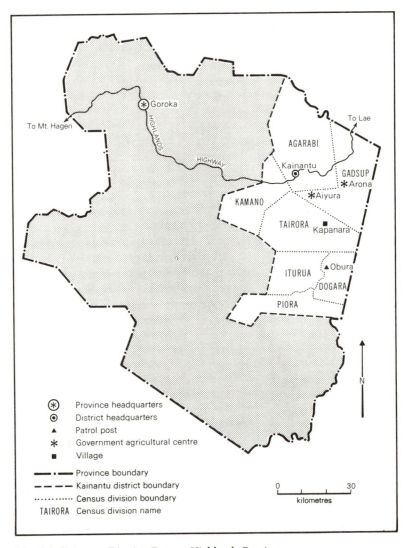

Map 1.2. Kainantu District, Eastern Highlands Province

of the Tairorə language speakers extends over several census divisions (see Map 1.3), though a major geographical and linguistic division between the Northern and Southern Tairora exists at approximately the southern border of the Tairora Census Division (Pataki-Schweizer 1980: 54).

The topography of the Northern Tairora region ranges from flat

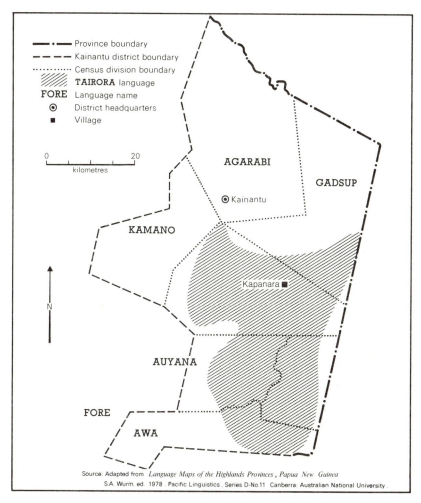

Map 1.3. Distribution of the Tairora Language

valley bottoms to undulating hills to small mountains. The area lacks the ruggedness characteristic of much of the Highlands. It is a mosaic of rain forest at various stages of succession, anthropogenic grasslands, and swamplands. Grasslands predominate. Each village territory has its own unique combination of these environmental zones. Situated in a valley in the northeast edge of the Kratke Range, Kapanara is well endowed in all three zones. Steep forest-covered hillsides form a basin rim around the northern, eastern, and western portions of the village boundary. Within the basin lie gently rolling grass-covered hills, the most extensive feature of the landscape. Narrow tree-lined watercourses traverse the open grasslands and form the northern tributary of the Lamari River system, which drains to the south coast. Intermixed within the grassland area are patches of swampland in the poorly drained valley bottom. The altitudinal range encompassing hamlets and most gardens, approximately 1,650 to 1,850 meters, is relatively narrow compared with some other Highlands areas.

Each environmental zone has special cultural significance. The forests are a repository of many edible resources such as nuts, leaves, insects, fungi, small mammals and birds, as well as a source of materials for house and fence building, firewood, cordage, and dyes. They also provide items used in sorcery and healing. The water here is cool and considered more refreshing than the warmer water of the grasslands. When people become enervated by the heat, they seek refuge in the cool forest for a few days. The forests also have many potential dangers for the unwary. Hidden sorcerers from other villages as well as malevolent spirits can harm those alone in the forest. The open grasslands are much safer but offer fewer resources. People obtain the grass *Imperata cylindrica* for roofing, some cordage, a few small edible mammals, and edible ferns from the grasslands. One of the most relished dietary supplements, the munah beetle (*Lepidiota vogeli*), is found here. The marshlands are important because they provide excellent foraging for the village pigs. Traditionally, people feared this zone most, and until recently large areas of swampland had not been cultivated within memory. Villagers believe that worms, or *ahe'u*, inhabiting the swamps possess evil powers.

The village is linked to the outside world by an unpaved road that bisects its territory. The road, which leads to Kainantu 26 kilometers away, is adequate during most of the year but can become quite treacherous for driving during periods of heavy rainfall. A sporadic trickle of traffic, which consists mostly of public motor vehicles (PMVs) and the trucks of coffee buyers, passes the village each day. The PMV operators, who are all nationals living in other villages, charged adult passengers 70 toea in 1977 for the one-hour ride to Kainantu. Despite the occasional availability of motorized transport, most people walk to nearby villages using either the main road or the traditional narrow footpaths.

Adjoining the road, the village hamlets are in the center of the grassland area. As is typical in the Northern Tairora area, the settlement pattern is nucleated (see Map 1.1). Residences in the hamlets are constructed of wooden frames, plaited bamboo walls, and grass roofs. Most are round with a conical roof, and the others have a rectangular floor plan. An inspection inside a house reveals a relatively simple technology and few material possessions. All households own a few metal cups, plates, utensils, pots, and pans. Some have a kerosene lamp, flashlight, and perhaps a few purchased tools such as a hammer or saw. Every household possesses a steel axe, bushknife, and spade, which are the major store-bought gardening implements. Traditionally prepared artifacts such as bow and arrows, string bags, and *Pandanus* leaf mats are ubiquitous. Most households own at least one store-bought blanket, and a few own pillows. Rough wooden beds, a table, and boxes may also be found. A few people own such luxury items as a radio, pressurized lamp, or umbrella. In most cases, that is all.

Interspersed within the hamlets are small, rectangular-shaped, family-run trade stores, which sell a variety of basic items such as imported rice, canned fish and meat, matches, soap, tobacco, kerosene, and paper for rolling cigarettes. With the closest town one hour away by vehicle, having locally operated trade stores makes it more convenient to purchase food and other items.

Central to the main cluster of hamlets is the *Eria Komuniti* (*Tok Pisin* for "Area Community") center, which has a fenced green as well as a jail and courthouse constructed largely of traditional materials. Introduced into the Kainantu District in the 1970s, *Eria Komunitis* are small, local governing units composed of two or more villages that voluntarily joined together to perform certain functions such as stimulating economic development, collecting taxes,[4] and establishing laws on matters of local interest such as bridewealth price and compensation for pig damage to gardens (Uyassi 1975). People gather occasionally at the *Eria* center to attend village court sessions, participate in *Eria* meetings, vote in elections, and hear talks by occasional government visitors on the importance of economic development, taxes, education, politics, and other matters.

Beyond Kapanara lie many similar Tairora villages where relatives, trading partners, and traditional enemies live. Most nearby villages were hostile before pacification when intervillage warfare

[4] The annual tax rate set by the Kapanara *Eria Komuniti* from 1976 to 1981 was K6.00 per male adult. This was the only direct tax levied on all households. Some who engaged in wage labor also had to pay an income tax to the national government.

was chronic. Although major warfare ceased in the late 1940s, the Kapanarans are still suspicious of many people from other villages who are perceived as the source of much-feared sorcery attacks. There are also several small plantations in the Tairora area, specializing largely in coffee production and some cattle holdings. The most important one for the Kapanarans is the coffee plantation at Noraikora, which in the mid-1970s employed a considerable number of people, both youths and adults, during the peak of the coffee-harvesting season.

SOCIAL ORGANIZATION

The people of Kapanara belong to one of three major named social groups: Kapanara, Iraerabura, and To'ukena. Within each group, patrilineal descent is the major stated criterion of recruitment, though some individuals who claim to be a member of a particular group cannot demonstrate agnatic links to others.[5] This emphasis on patriliny is characteristic of Tairora villages (Watson 1970: 109), though in certain contexts other types of relationships, such as cognatic descent, may be important.

Figure 1.1 provides a highly abbreviated outline of the main components of one of the largest social groups, To'ukena. Similar structures characterize the other two groups. Collateral lines have been ignored where possible for the sake of brevity. The youngest

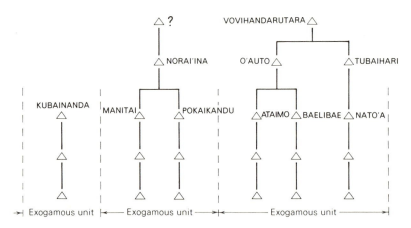

Figure 1.1. To'ukena Patrilineage Cluster

[5] In some cases, individuals become members of a group not by descent but by cumulative patrifiliation (see Barnes 1962).

generation denotes present-day adults. I call each group a "pat-rilineage cluster"[6] because each is composed of different patrilin-eages lacking a common ancestor.

When referring to social groups within their patrilineage clus-ter, adults often use the name of their agnatic grandfather. Thus, members of the line of Pokaikandu (see Figure 1.1) distinguish themselves from those of the lineage of Ataimo. These smaller "minimal patrilineage units" are significant in relation to inter-group competition and in recruitment to *bisnis* enterprises.

Intermarriage occurs within and between the clusters. As is true in other Tairora villages (Watson 1967: 59), a majority of men and women marry within their own village, though no prescrip-tive rule of village endogamy exists. The pattern of marrying within the village has important implications for the flow of goods and services because many important exchanges of food and money are conducted among affines.

The married couple plus dependent children, if any, form the basic economic unit, the household.[7] Monogamous households predominate, though polygynous households and incomplete ones headed by a single adult also exist (see Table 1.1). Most households live in their own separate dwelling, though some share a single house. Members of polygynous households live together unless considerable antagonism exists between co-wives.

SETTLEMENT PATTERN

Since the road linking Kapanara to Kainantu was constructed, most people have preferred to live near or next to it (see Map

[6] I am indebted to Dr. Marie Reay for suggesting this term.

[7] Because of potential confusion over the use of the term "household," a note of clarification is necessary. No household is totally independent in production and consumption. All households share food with others and help others occa-sionally in certain productive tasks. The degree to which units are independent determines whether they are considered households. Some adults are either wid-owed or divorced. Whether these unmarried adults plus dependent children are designated as separate, "incomplete" households depends on whether they are relatively independent economic units. Thus, single elderly adults are considered as separate, incomplete households if they produce most of the food they consume. If weak and dependent, they are counted as belonging to the household of the person supporting them. As the elderly become older, there is a gradual transition from relative independence to dependence, and in some cases classification was difficult. I also consider polygynous families to be one household. Even though each co-wife is a relatively independent producing unit, the entire family is or-ganized by a single man and is thus classified as one household.

TABLE 1.1

Frequency Distribution of Household Types, 1977

Type	Total Number	Percentage of Total
Monogamous	75	70
Polygynous with two wives	13	12
Polygynous with three wives	4	4
Polygynous with four wives	1	1
Total Married	<u>93</u>	<u>87</u>
Incomplete, widowed	11	10
Incomplete, divorced	3	3
Total Incomplete	<u>14</u>	<u>13</u>
Total Households	<u>107</u>	<u>100</u>

1.4).[8] The road is part of the communication network linking the village to the outside world, in general, and to government personnel, in particular, who provide administrative, health care, and advisory services; these agents often disseminate news while passing through the village on the road. The road is also the focal point of many activities such as selling coffee or playing cards. Watching others sell coffee or awaiting the return of fellow villagers from town often provide the occasion for gatherings on the road. Relative accessibility to PMVs passing through the village is another advantage because people enjoy travelling to town, especially during the coffee-harvesting season when more money is available. Roads have been an important force in attracting hamlets in many other areas in the Highlands (see Sorenson 1972: 366).

The road is not the only influence on settlement pattern. People also want to be near their coffee plots, where they spend much

[8] In approximately 1945, an administration police officer settled the people into one large village in the valley floor in open grassland. In 1950, after their last major battle, the villagers abandoned the site in order to move farther away from the enemy and separated into two nucleated hamlets according to patrilineage cluster affiliation. Since that time, the number of hamlets in the village gradually increased. However, in 1964 a villager who had served the administration in various capacities settled most of the people again into one large nucleated village so that he could more easily disseminate news, especially that related to the forthcoming elections for the first House of Assembly. After a few years, the number of hamlets again gradually increased, until in 1977 there were six hamlets.

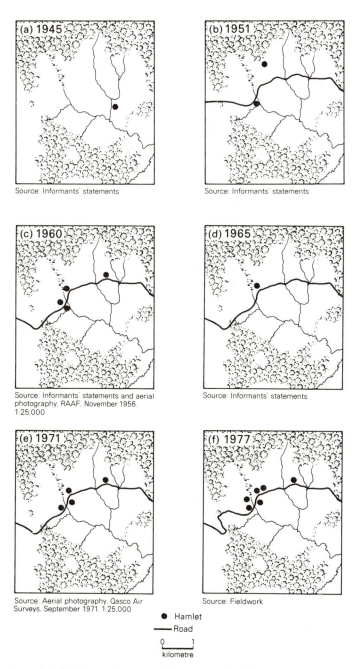

Source: Informants' statements

Source: Informants' statements

Source: Informants' statements and aerial
photography, RAAF, November 1956.
1:25,000

Source: Informants' statements

Source: Aerial photography, Qasco Air
Surveys, September 1971, 1:25,000

Source: Fieldwork

● Hamlet
── Road
0 ___ 1
kilometre

Map 1.4. Settlement Changes in Kapanara, 1945-1977

time during the coffee flush from May to August harvesting the major portion of their crop. They also need to guard against theft, a major problem in the village, as is evidenced by the numerous complaints about stealing coffee during the flush. By planting coffee near the hamlets, guarding against theft is easier. The need to protect coffee, in turn, places a constraint on moving too far from one's plots.

The location of the road and coffee set major constraints on residential movement in the sense that people do not want to move too far from them; new hamlets, for example, are established in close proximity to older ones. Within the zone in which people feel the distance to their coffee and the road is tolerable, other factors motivate residence changes.[9] Some people, after living in the same place for perhaps ten years or longer, tire of the site and want a change. Others move from the center of the hamlets to get away from the pigs, dogs, children, and family quarrels because they prefer a measure of solitude. Occasionally, bitter quarrels, whether over money, women, food, or pigs, lead to changes in residence, in which case the particular aggrieved person may burn down his home in anger. At other times, a kinsman or affine invites an individual to build a house next to his. A few people moved because they were living near the coffee groves and feared that sorcerers could easily hide there and attack them. When a new hamlet is being formed, some people move simply because others have done so. Serious illness and death can also influence residence changes. For example, in 1957 an epidemic, which was most likely influenza (A. Vincent, personal communication 1977), caused the deaths of several people in one hamlet; because sorcery was suspected as the cause, the entire hamlet was burned down, and the residents moved elsewhere. Accessibility to garden land, although an important consideration, historically has not been a major determinant of settlement pattern because of the abundance of land and the flexible system of land tenure (cf. Brown and Brookfield 1967; Waddell 1972).

Cattle projects, which are also located next to the road (see Map 1.5), have become a significant influence on residential movements. Agricultural extension officers, who are based in Kainantu and supervise the smallholder cattle projects, have successfully urged most project leaders to reside within or next to their project,

---

[9] People have to build a new house every three or four years, though many often build near or next to their old house.

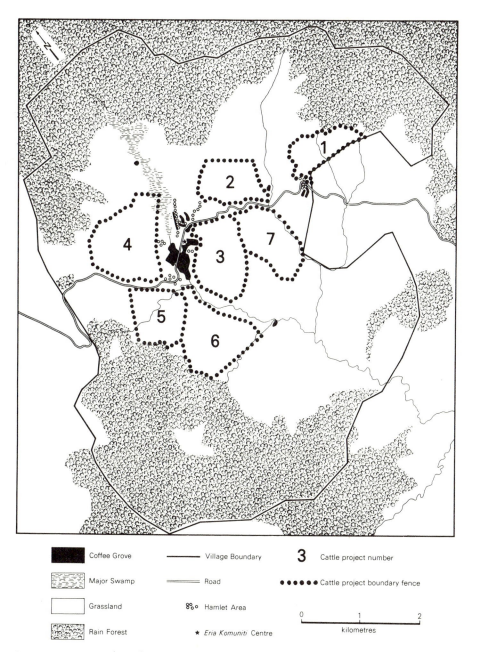

Map 1.5. Location of Cattle Projects in Kapanara

hoping that better care will be provided for the cattle. Several households involved in one enterprise established a new hamlet just outside their project in 1975 to symbolize their control over the land. In addition, because villagers do not establish new hamlets within the land fenced for cattle projects, their ability to make future changes of residence within the vicinity of the coffee groves and the road is further constrained.

POPULATION

Population density in relation to arable land is a key variable affecting commercial development. Where densities are very high, most peasants have access to only small plots to farm, and consequently they live a precarious existence. The limited land availability means that in order to increase cash cropping, subsistence production has to be curtailed. Facing considerable risk in allocating resources to commodity production, such peasants are likely to hesitate in significantly increasing their commercial output unless forced to do so by external forces. In contrast, the Kapanarans, having considerable land, are not constrained by these problems.

Kapanara is one of the larger villages in the Northern Tairora area, having a population of 441 in 1976.[10] As the village territory is 39 square kilometers, the population density is 11 per square kilometer, which is low for the Highlands. From the perspective of many development planners, a situation of such abundant land resources provides a prime opportunity for commercial development.

Two features of the age-sex structure of the Kapanara population are noteworthy (see Figure 1.2). First, the shape of the population pyramid suggests a potential for sustained population growth, though not at the explosive rate experienced in some underdeveloped countries. Thirty-nine percent of the population is 14 or under and constitutes the future potential reproductive pool. Second, the age-sex structure is not significantly affected by wage-labor migration today. In 1976, for example, only six adult males were away from the District. With a low population density, everyone still had access to adequate land for farming and cash cropping; in other regions in the Highlands where high population densities exist, resulting land shortages have stimulated considerable male wage-labor migration. Previously, when men

[10] The population of Northern Tairora villages in 1980 ranged from 127 to 448.

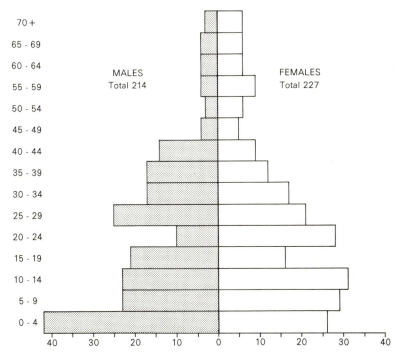

Figure 1.2. Population Structure of Kapanara, July 1976

worked as contract laborers on the coast under the Highland Labour Scheme from the mid-1950s to 1974, many Kapanaran adult
males were absent from the village, but at that time cash-earning
opportunities at home were extremely limited.

Data on population growth for the Northern Tairora are available from 1958 to 1972, during which the population grew from
4,925 to 7,223 (PNG, Kainantu District Headquarters files), an
annual rate of population growth of 2.77 percent. For the same
period, the population in Kapanara rose from 296 to 414, an annual
growth rate of 2.43 percent, which is slightly below the Northern
Tairora average. With the abundance of land within the territory,
no problem of *total* population pressure is foreseeable. However,
the population will not be uniformly distributed over the land
but concentrated near the prime areas, presently defined mainly
by the road and the coffee plots, and thus pressure on the most-
desired areas will increase substantially. In addition, cash-earning
activities such as cattle raising, which utilize substantial amounts
of land, will exacerbate the problem.

CHAPTER TWO

Smallholder Cattle Projects

WHEREAS EUROPEANS have been raising cattle in Papua New Guinea since the late 1800s, villagers did not become significantly involved in the industry until the 1960s. Numerous considerations, both local and national, influenced the government to stimulate village smallholder development. In turn, villagers became eager to start cattle enterprises, though their motivations differed from those of the government. Changing government policies concerning cattle raising have radically altered the external, political-economic environment, increased the potential for conflicts between subsistence and commodity production, and greatly facilitated rural economic differentiation.

THE HISTORY OF THE CATTLE INDUSTRY IN PAPUA NEW GUINEA[1]

Although cattle are found in villages throughout Papua New Guinea today, they are not native to the country. In the late nineteenth century, German colonists living on the northern coast introduced cattle, and by 1900, European settlers owned 250 head (Purdy 1972: 138). They kept most cattle on lowland coconut plantations, where the animals served the dual purpose of keeping down weed regrowth and of providing fresh meat and dairy produce for the plantation personnel. Missionaries also raised a few cattle for their own station's use. By the outbreak of World War II in 1939, the number of cattle in Papua New Guinea[2] had grown to 39,100 (Commonwealth of Australia 1940a,b), but most were eaten during the Second World War, and the next available census

---

[1] I obtained part of the data used in this section during my tenure as a Visiting Research Fellow at the Papua New Guinea Institute of Applied Social and Economic Research, March-April 1979. See Grossman (1980) for a more detailed history of the cattle industry.

[2] Formerly the Territory of Papua and New Guinea, the country became known as Papua New Guinea in 1971. The Australian government granted Papua New Guinea independence in 1975.

in 1951 indicated that there were only 3,700 head (Commonwealth of Australia 1952a,b).

In the late 1940s and early 1950s, the Australian administration became concerned over growing beef imports, and it thus decided to increase the number of cattle in the country to reduce dependence on overseas sources and to eventually become self-sufficient in beef. The administration became convinced that a viable cattle industry run by private enterprise was needed, and to achieve its goal, the administration took three steps in the mid-1950s to encourage the growth of the beef-raising industry. It granted several large pastoral leases to Europeans, provided free veterinary and advisory services, and introduced a subsidy scheme to lower the high cost of transporting cattle from Australia to Papua New Guinea. The basis was thus established for the development of the beef cattle industry in the European private sector.

Through the 1950s, cattle raising was almost exclusively a European activity. Villagers owned only a few animals, obtained mostly from local missionaries. In the early 1950s, however, the administration began discussing the possibility of actively encouraging village cattle projects (Commonwealth of Australia 1953: 33). Several goals motivated the Department of Agriculture, Stock and Fisheries (DASF), the administration agency responsible for cattle development, to assist villagers in starting cattle projects. One was to use large areas of available grassland where population densities were low (McKillop 1965: 11). Another was to increase the supply of animal protein in the village diet, particularly in the Highlands, where the intake of animal protein was low (Anderson 1963: 173). In addition, in the late 1950s, DASF agricultural extension officers felt that nationals were planting too much coffee in view of the growing world surplus and that cattle projects would be an acceptable income-earning substitute (Finney 1973: 65-66).

DASF officially introduced smallholder cattle projects when it established three village projects near Goroka in 1960 and seven more throughout the Eastern Highlands in 1962 (McKillop 1965). Villagers had to purchase their own heifers,[3] but a grant from the Reserve Bank of Australia provided free fencing, pasture seed, and yard materials. DASF subsequently helped start other self-fi-

[3] Cattle on smallholder projects and European-owned ranches are crossbreeds between *Bos indicus* (Zebu breed) and *Bos taurus* (British breed). Most cattle have between one-half and three-quarters Zebu "blood."

nanced projects throughout the country, though they occasionally made informal credit arrangements to facilitate the purchase of fencing material (McKillop n.d.: 7). Extension officers provided free veterinary services, advice, and subsidized transport to assist the new cattle owners. The chief of the Division of Animal Industry stated optimistically (Anderson 1963: 172):

> The introduction of cattle to the New Guinean's agricultural system will be a great change from the present pattern and should be of profound importance to the New Guinean's development.

These early village breeding and beef-fattening[4] projects were quite small in most areas. DASF policy stipulated that at least four heifers had to be purchased initially (McKillop 1965: 13), and most projects had less than ten head altogether (Clark 1970: 4). Because of the small size of the projects, low prices received from the sale of steers, and inefficient management, the economic returns from these early projects were poor, and the failure rate was high.

Projects adopted a day-herding/night-paddock system. Villagers had to erect a small, fenced holding paddock of approximately two to four hectares and plant some improved pasture within it. During the day they herded the cattle outside on fallow grassland and at night confined them inside the fenced paddock. The rationale for the day-herding/night-paddock system was based on DASF's understanding of the traditional land tenure system: the grasslands belonged to the community as a whole, but only a few individuals supposedly owned the cattle, and thus the rest of the community would object strongly to enclosing a considerable portion of the grasslands for raising cattle (Anderson 1963: 171). Allowing cattle to graze outside the paddock during the day would present no problem, DASF believed, because the rights of others to use the grasslands would not be affected. With only a small area enclosed, the system would not conflict with subsistence production. The administration also intended that the night paddocks would be used as a future garden site because the build-up of manure would make the soil very fertile. Thus, DASF specifically wanted to integrate cattle grazing and gardening, hoping to

---

[4] Smallholder cattle projects have always remained beef-fattening and breeding enterprises. None are dairy farms.

avoid a conflict between subsistence and commodity production. The chief of the Division of Animal Industry (*ibid.*) asserted:

The final agricultural system under which the New Guinean will live *must* [emphasis in original] be one of mixed farming where an animal must enter into the rotation of the agricultural land, both to increase fertility of the fallow land and to turn into profit the crops grown in this fallow.

Although the system may have seemed ideal, it did not work well in practice. The smallholders' level of management and knowledge of cattle husbandry were poor by European standards. Villagers did not always properly herd cattle during the daytime. Occasionally, they allowed cattle to congregate in areas of good pasture, which became overgrazed and infested with weeds (*ibid.*). Sometimes they left them unattended, and the animals roamed through the village destroying gardens. In more extreme cases, people left cattle in the small night paddock for several days, and overgrazing resulted (Clark 1970: 5). Confining cattle to a small paddock for several days facilitated the rapid build-up of the number of debilitating internal parasites. The outcome was discouraging; cattle lost condition, and high mortality rates resulted.

In order to alleviate these problems, DASF changed its policy. By 1970, the day-herding/night-paddock system was abandoned in most places, and in all subsequent cattle projects, villagers had to enclose all the grazing land within a barbed wire fence, a policy that in effect removes a considerable amount of land from the agricultural system. The original concept of mixed farming, integrating gardening and cattle husbandry, was thus abandoned, and the potential for conflict between subsistence and commodity production increased substantially.

DASF policies were not the only forces affecting smallholders. International development agencies have also influenced the development of the cattle industry. In 1963, while the smallholder cattle sector was still in its infancy, the Australian administration invited the International Bank for Reconstruction and Development (IBRD) to suggest appropriate strategies for development planning in Papua New Guinea. The IBRD suggested that in a country with limited financial resources and manpower, the development effort "should be concentrated in areas and on activities where the prospective return is highest" (IBRD 1965: 35). Viewing the cattle industry as an activity with such potential, the IBRD recommended a major expansion.

The IBRD Mission envisioned a two-stage program. In the first stage, the herds on administration and European ranches would be increased and more pastoral leases made available. Consistent with its "concentration of effort" principle, the IBRD emphasized the importance of focusing initially on the large-scale ranching sector to utilize the Europeans' skills and experience in more efficiently increasing the size of the country's herd (*ibid.*: 136). In the second stage, the breeders and calves produced on administration and European ranches would stock the rapidly growing smallholder sector.

Responding to the suggestion made by the Mission, the administration established the Papua New Guinea Development Bank (PNGDB) in 1967. Previously, a key factor limiting economic growth in general and the expansion of the livestock industry in particular was the lack of available credit. Villagers' sources of credit for establishing commercial enterprises were especially limited. Since its opening, the Bank has provided many loans for smallholder cattle projects.

The cattle industry received additional support when the International Development Association (IDA), an affiliate of the World Bank, provided two loans to the PNGDB to expand the ranch and smallholder sectors respectively, in accordance with the two stages envisioned by the IBRD. With this funding, the PNGDB was able to increase its lending to the cattle industry. In 1969, the first loan provided a credit of $A1,400,000 to develop the large-scale ranching sector. Expatriates received most of the loans for private ranches, with only a few nationals obtaining assistance. In 1973, the second IDA credit provided $A4,200,000 for the development of smallholder livestock projects in beef, pigs, and poultry. Most of the credit, $A3,500,000, was initially allocated to finance the stocking of smallholder cattle projects and a further $A338,000 for purchasing equipment, improving extension services, and expanding smallholder and in-service training and research related to improving the smallholder cattle industry.

The PNGDB's entry into the smallholder cattle sector radically changed the industry's nature. Previously the number of stock on a village project could be increased significantly only on a "pay-as-you-go" basis. With the availability of credit, much larger projects could be established immediately, though often cattlemen were first required to purchase their own barbed wire and erect a boundary fence before obtaining a loan. The PNGDB commonly approved loans ranging from K2,000 to K5,000, sums far beyond the annual rural household cash income. From 1967 to December 1977, the PNGDB granted 3,195 loans for smallholder projects for a total value of K7,013,500 or 41 percent of all agricultural loans to nationals (PNGDB 1976/77, 1977). Of the 3,300 projects functioning in the country in 1978 (DPI files, Port Moresby), the PNGDB funded approximately 2,600, assuming that roughly 600 of the 3,195 projects that received loans were defunct by 1978. The other 700 functioning smallholder cattle projects were self-financed (Grossman 1980).

The entry of the PNGDB into the nationally owned sector also established a new social relationship between a large bureaucracy and the villagers. Such government credit institutions are common throughout the Third World and generally have a profound impact on rural communities. The policies and regulations of the PNGDB have affected village land-use patterns, the extent of rural economic differentiation, and the level of community autonomy.

The size of the smallholder herd grew rapidly after 1967, reflecting the influence of the PNGDB (see Figure 2.1). The number of cattle owned by nationals rose from 3,635 in 1967 to approximately 50,000 in 1978, or from 8 percent of the country's herd to 37 percent (PNG, Bureau of Statistics 1970/71; DPI files, Port Moresby). Villagers established projects throughout the mainland, with those in the Eastern Highlands Province, an area endowed with undulating grasslands and low population densities, having the most smallholder enterprises in the Highlands region (see Map 2.1).

As the number of smallholder projects increased, agricultural extension officers, who are the PNGDB's field agents, devoted more of their time to supervising the projects at the expense of other forms of rural economic activity. McKillop (1975: 17) reported that Rural Development Officers, the most highly qualified extension officers, spent 59 percent of their contact time with smallholder cattlemen. Thus, a large percentage of the extension effort benefited only the cattle owners, a very small minority of the rural population (ibid.). This imbalance was slowly being corrected in the late 1970s and early 1980s as less official emphasis was being placed on the expansion of the smallholder cattle sector.

The rate of growth in the smallholder sector has decreased significantly since 1975 (see Figure 2.1). As an indication of the slower growth rate, the annual number and amount of PNGDB loans to national cattlemen have declined appreciably since 1973-74 (see Figure 2.2). Several factors account for the slowdown in

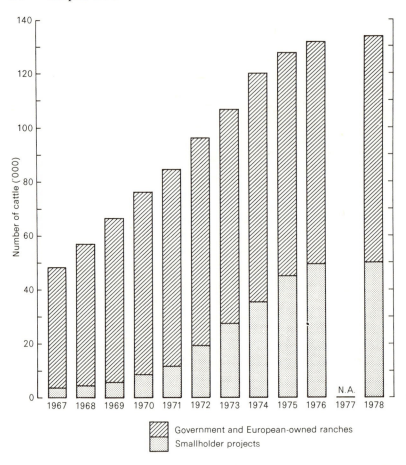

Figure 2.1. Growth of the Cattle Industry in Papua New Guinea, 1967-1978. *Source*: Bureau of Statistics 1970/71, 1975/76; Cairns 1976; DPI files, Port Moresby; Grossman 1980.

the growth rate of the smallholder herd. In many areas, villagers have already fenced off much land for cattle, and enclosing additional land would cause further strains in the subsistence system. Their poor level of management results in low calving rates and high mortality rates. The calving rate in the smallholder sector is only 53 percent (DPI files, Port Moresby), whereas on European-owned ranches it is approximately 70 percent (Densley n.d.: 9). The demanding repayment schedules of the PNGDB loans often dampen the enthusiasm of many involved in cattle projects; because of short repayment periods of five to nine years stipulated

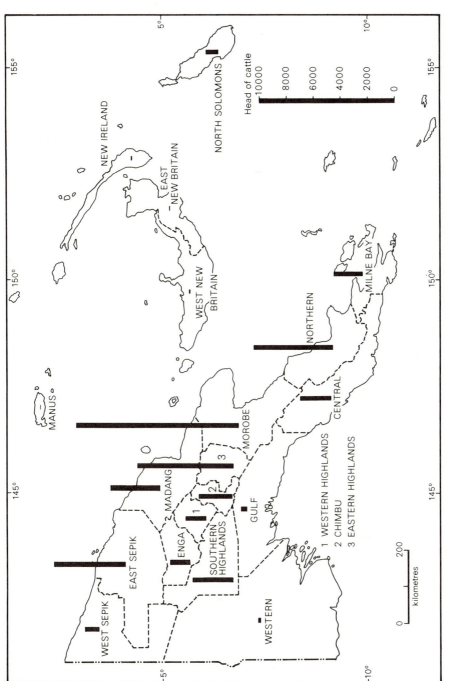

Map 2.1. Distribution of Cattle Owned by Smallholders in Papua New Guinea, 1976.
Source: Bureau of Statistics 1975/76; Grossman 1980.

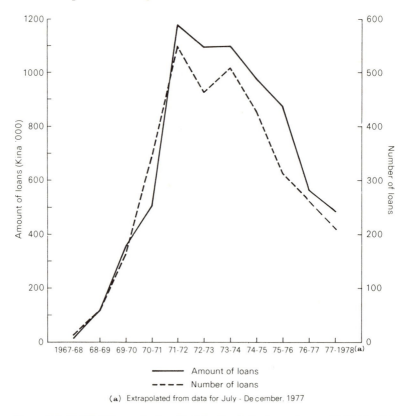

Figure 2.2. PNGDB Loans for Smallholder Cattle Projects. *Source*: PNGDB 1974/ 75, 1976/77, 1977; Grossman 1980.

in the loan agreement, most revenues produced in the first few years of a project's operation are supposed to be sent to the PNGDB, leaving little for distribution within the village. Changes within the agricultural extension staff are also slowing the development of the smallholder sector. The percentage of national staff increased markedly in the mid-1970s. Unlike European extension officers, who vigorously promoted cattle projects, national officers prefer to let the initiative for cattle development originate with the villagers themselves, and thus extension inputs into cattle raising are decreasing, particularly in the early 1980s.

This variation in the extent of extension inputs over time into the smallholder cattle sector is particularly significant. If we expect that the spread of commodity production can negatively affect subsistence production, and if the spread is facilitated partly by extension activity, then a decline in such extension inputs might possibly be a benefit to subsistence production. This issue is, in fact, rather complex and can only be explored fully by examining the changing impact of the commercial economy at the local level.

SMALLHOLDER CATTLE DEVELOPMENT IN THE KAINANTU DISTRICT

As in other areas of Papua New Guinea, the early cattle projects in the Kainantu District employing the day-herding/night-paddock system had a high failure rate, and the number of cattle in the projects grew slowly. In 1968, the year before the PNGDB provided loans for smallholder cattle development in the Kainantu District, villagers owned only 188 head of cattle on 33 projects (Clark 1970: 5), with 84 percent of the enterprises having 10 head or less.

The provision of credit from the PNGDB was clearly the decisive factor stimulating the growth of the smallholder cattle sector in the Kainantu District (see Figure 2.3).[5] Nineteen-seventy was the first year in which a significant number of projects were stocked using PNGDB credit, and rapid growth ensued immediately. Enterprises with PNGDB loans were started throughout the District, with most located in the southern and eastern portions, where population densities are lowest (see Map 2.2).

In terms of the IBRD's recommended principle of concentration of effort "[in] large areas of good land which are relatively accessible and where development is relatively easy" (IBRD 1965: 35), the Kainantu District was a prime area for cattle development because it has low population densities, mild slopes, and extensive grasslands. The area is serviced by the Highlands Highway, the main transport artery linking the Highlands to the coast, and the Kainantu District itself contains a relatively well-developed road system that reaches 95 percent of the villages.

Other factors also influenced this rapid growth. Animal husbandry is an integral part of Highlands culture, but Seventh-Day Adventists, who are numerous in the Agarabi and Kamano Census Divisions in the District, are forbidden to raise pigs. Adherents

[5] Because of the irregular nature of data collection in the District, the data do not accurately reflect annual changes though they are useful to illustrate growth over several years.

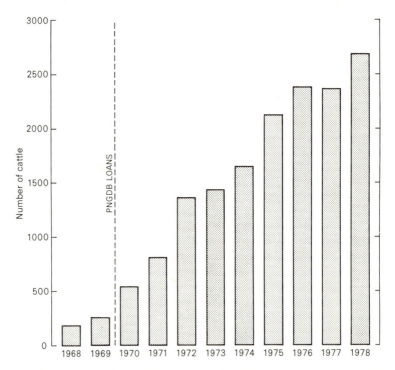

Figure 2.3. Growth of the Smallholder Cattle Herd, Kainantu District, 1968-1978.
Source: Clark 1970; DPI files, Kainantu and Port Moresby.

to the faith found cattle an acceptable substitute. In addition, European agricultural extension officers vigorously promoted smallholder projects, encouraged villagers to apply for loans, and often reassured them that the financial returns would be considerable. Although many dedicated extension officers believed that cattle projects were an appropriate form of development to improve rural living standards, other extension agents promoted projects because they perceived their own potential for advancement within the government bureaucracy as being dependent partly on the number of new cattle projects the agent helped start in his area. Also, indigenous demand for cattle projects was considerable, especially because the only major village cash-earning alternative was coffee production, for which prices were very low in the early 1970s. Everyone seemed optimistic about the future of the smallholder cattle industry in the Kainantu District.

Extension officers were eager for the cattle projects to succeed, and to further their goal, they helped form the Kainantu Bulu-

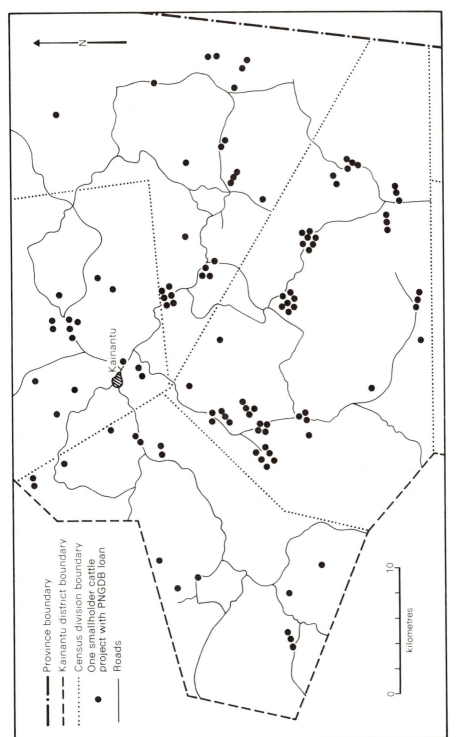

Map 2.2. Distribution of Smallholder Cattle Projects with a PNGDB Loan in Kainantu District, 1976. *Source:* DPI files, Kainantu.

makau Association in 1972. The aims of the organization were to provide a forum for cattle owners to meet on a regular basis to facilitate the extension officers' efforts in disseminating information concerning the industry, to enable cattle owners to discuss their joint problems as cattle husbandmen, and to develop policies on matters affecting the smallholders.

In the first few years of operation, the Association met regularly in Kainantu, but in subsequent years meetings have had a more erratic schedule. Extension officers tried to decentralize the organization by having cattlemen meet in their own census divisions, but their efforts were not successful. Cattlemen viewed themselves as men of *bisnis* whose status should have appropriate recognition. Members preferred to be picked up at their home village by extension officers and be driven to the town along with other cattle project leaders. These larger meetings in town were held in the local government council chambers, a relatively expensive and impressive structure. The desire to have transport provided and to attend meetings in relatively lavish halls in town centers is a common characteristic of many other rural interest groups in Papua New Guinea and is a reflection of their elitist consciousness (Gerritsen 1979). Other villagers rarely received such help from extension agents and hardly ever used the town hall.

The number of members who attended the meetings in town varied between 50 and 100, but usually no more than 5 or 10 of the more articulate and demonstrative individuals dominated the discussion. People talked about such issues as cattle prices, problems of pigs digging up pastures, the high price of veterinary supplies, bovine diseases, difficulties in transporting cattle, and a host of other topics. Permanent membership dues were only K2.00, but many who attended were not dues-paying members.

Extension officers advising the organization tried to instill in the members a sense of identity and pride in their cattle enterprises. In 1973 and 1974, symbols of group unity were adopted in the form of badges and T-shirts bearing an emblem and the name of the Association. Members were told, "Badges should be worn at all times, thus showing that the wearers are cattle businessmen" (PNG, Kainantu Bulumakau Association Minutes, June 1973). Smallholders frequently wore their Association T-shirts when visiting town. Although extension officers stated that they were not attempting to instill a sense of superiority in the cattlemen, the effect of their actions often resulted in the cattlemen believing that they had a certain status recognized by government officials that other nationals lacked.

This pattern of extension activity—spending a lot of time with cattle smallholders, fostering their self-image, and establishing the Kainantu Bulumakau Association—has created what appears to other villagers as a privileged relationship between extension officers and cattlemen. Smallholders have been adept at manipulating this relationship to further their own economic interests at the expense of their fellow villagers.

The 1978 nationwide census of smallholder cattle projects provides the latest available data for the Kainantu District (DPI files, Port Moresby). In 1978, villagers owned an estimated 2,681 cattle on 164 projects, with a mean of 16.4 cattle per project (see Figure 2.4). A majority of the projects were financed by loans from the PNGDB, and these generally had more cattle than self-financed enterprises. The average project was approximately 60 hectares in area. As the term "smallholder" suggests, the number of stock owned and the average area enclosed were relatively small compared with the much larger European-owned ranches, some of which were over 5,000 hectares in area and had over 5,000 head of cattle in the mid-1970s (Bishop and Redish 1977: 1). Nevertheless, as the story of Kapanara reveals, the impact of cattle raising on rural communities can be substantial.

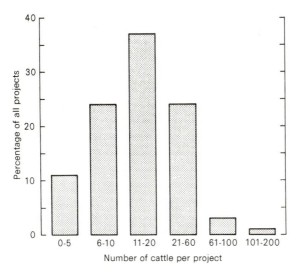

Figure 2.4. Frequency Distribution of Number of Cattle per Project, Kainantu District, 1978. *Source*: DPI files, Port Moresby.

CATTLE IN KAPANARA

When the PNGDB was established in 1967, the people of Kapanara had little understanding of the outside world in general or of cattle husbandry in particular. Their first sustained contact with a cattle project resulted from their selling the administration in 1968 a 260-hectare block of land at the southeastern boundary of the village territory (see Map 2.3). In 1969, Bonafo, a national from the far west corner of the Eastern Highlands Province, leased the land from the administration for a cattle project and obtained over 100 cattle with the aid of a PNGDB loan. Influenced by Bonafo's statements that the cattle business was very lucrative, villagers in 1970 built their first project, which adjoined Bonafo's. In 1973, they constructed two more, and in early 1975 established another four. In 1977, there were 162 head of cattle in the seven projects, which enclosed a total area of 5.49 square kilometers of grassland (see Table 2.1).

Although the patterns followed in the establishment of each project varied slightly, certain fairly standardized steps from inception to completion can be recognized. Similarly, relationships between villagers on the one hand and the agricultural extension officers and the PNGDB on the other also exhibit certain regularities.

Once a man of some esteem—a big-man, a successful entrepreneur, or a more dynamic member of one of the patrilineages—decides to enter the cattle business, he calls together people related to him through various ties of kinship, descent, and friendship, seeks their support to start a cattle project, and suggests that they use a particular parcel of land to which they have a claim. The man who has called the others to the meeting acts as a leader or "cattle boss" of the project; the others will be his "followers" or "helpers." At the same time, the leader usually seeks another person of some repute to join him in the endeavor. This person may be chosen for a variety of reasons, none of which are mutually exclusive. He may have saved money from cash-earning activities or he may be a close relative. In addition, he or his kinsmen may have a major claim to the land to be enclosed. This person in turn recruits his own relatives and friends, discusses his intentions, and acts as the leader or cattle boss of the second group. Thus, two groups are usually joined together.

A cattle boss attempts to recruit a following that is large enough to obtain the necessary land, labor, and finance to start the project.

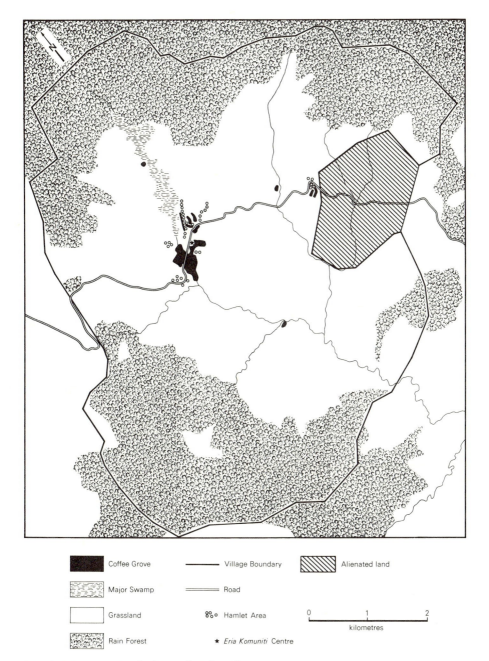

Map 2.3. Kapanara Land Alienated to the Administration

TABLE 2.1

Characteristics of Cattle Projects in Kapanara, November 1977

Project Number	Year Constructed	Number of Cattle	Area Enclosed (ha)
1	1970	26	48.1
2	1973	20	60.4
3	1973	44	85.3
4	1975	22	107.1
5	1975	25	70.4
6	1975	16	100.6
7	1975	9	76.6
		162	548.5
		Mean = 23	Mean = 78.4

Usually, when a leader recruits a man or woman, the other spouse of the household also joins. Normally, a cattle boss actively recruits followers, though a few individuals have successfully sought permission to join a project. Agnation is the principal tie used in recruitment; fifty percent of the households involved in projects are headed by men of the same minimal patrilineage unit as their project leader. The pattern of recruitment reflects the strong solidarity among members of the minimal patrilineage unit. Most land enclosed is part of the estate of the cattle boss's minimal patrilineage unit, and thus his agnates in the unit are joining an enterprise using their land. Another 17 percent of the households recruited are headed by affines of the project leader. Normally, a man will not refuse a request by an affine to join the latter's project even though the man prefers to join that of his own minimal patrilineage unit. The remainder of those recruited are linked to the project leaders by a variety of relationships. The number of individuals (counting both husband and wife) who joined to help establish each project ranged from 16 to 43.

In most cases, an individual will join only one project because participating in two obligates the person to contribute labor and money to both projects, an expenditure that is felt to be too demanding on financial resources and time. If a man joins his affine's project, he has no rights in the cattle enterprise established by others of his own minimal patrilineage unit unless he joins the latter project as well.

The cattle bosses appeal to their followers by proclaiming that the cattle project will enhance the prestige of their group, will earn them a considerable income, and will provide a locally available source of fresh beef. The leaders stress the advantages that will accrue to the group as a whole, though the exact distribution of future rewards is left unstated. The helpers expect to benefit substantially because they are giving up part of their land and will contribute labor and cash. A consensus is usually reached to start a project.

Next, the cattle bosses seek wider support for the proposed enclosure. They request the permission of the elders of their patrilineage cluster to use the land. The elders, though no longer leaders, are still considered, in a symbolic sense, guardians of the land. In a few cases, the cattle bosses invite all villagers to hear the proposed plan. Discussions ensue, often with complaints from those who object that either their present gardens or their fallow land will be enclosed. At these meetings, unanimity in favor of the business venture is not always reached, but as long as the opposition does not appear to be overwhelming, the project leaders decide to proceed. Then, the bosses inform the agricultural extension officers stationed in Kainantu that they want to start a project and obtain a PNGDB loan. The loan is needed to finance the enterprise because the villagers do not have enough cash to start an economically viable project.

The bosses return to the village and place tall bamboo posts at the corners of the proposed fence line to show clearly the project boundaries. This demarcation provides more concrete clues to the other villagers as to the project leaders' intentions than did the informal discussions that preceded the placing of the markers. It also provides the context for more substantive reactions. In a few instances, those who object strongly to the proposed boundaries cut down the bamboo markers or move them. Others simply voice their objections again. The spatial distribution of individual land rights is highly intermixed, and fencing off any large area is bound to impinge on the rights of some people. In a few instances, the project leaders defer to certain objections and change the proposed boundaries, whereas in others they ignore them and proceed.

The task of constructing a project is extremely arduous and time-consuming. Although some tasks involved are new, traditional labor patterns are followed. A sexual division of labor is practiced, based on established roles in gardening work and on the preconceived notions concerning the abilities and weaknesses of the two sexes.

Men go to the forest to prepare fence posts. The preferred wood for posts for both garden and cattle fences is the southern beech *Nothofagus* sp., which has a reputation for durability and strength. The men select tall, straight trees, fell and then cut them into sections approximately 2.3 meters long. These sections are then cut lengthwise into smaller and smaller portions using a combination of wooden wedges and levers to produce the appropriate-sized posts. The men strip away the outer bark and taper the top of the post so that rain will wash off quickly, preventing the post from rotting prematurely. The task of preparing the posts is entirely the province of men, who work in groups of five to fifteen. Women are thought to lack the strength and skill necessary for the task.

Soon after the men begin work, women arrive to carry the posts from the forest to the forest-grassland boundary. At the end of the workday the men and women gather together near the piles of cut fence posts to discuss the day's events and future tasks. By now, some women have harvested food from their gardens, and nearby they prepare a large earth oven, or *mumu* in *Tok Pisin*, to cook it. After eating, people sometimes return to the village in late afternoon. However, when the last four cattle projects were built, the men slept in the forest to save time travelling to and from the village.

The next step is to carry the posts to various points around the proposed cattle fence, a task that must be done when the grasslands are sufficiently dry, lest the people slip and injure themselves. A common stereotype concerning labor patterns in Papua New Guinea is that men do the physically demanding work, whereas women perform the less demanding but more routine and monotonous tasks. However, women usually carry two posts and on some occasions even three, but the men invariably carry only one. As an average post weighs approximately 16 kilograms, the effort is considerable. Men acknowledge the superior ability of women to balance and carry heavy loads and expect them to do so. They are quick to criticize if the women appear to slacken their efforts, which sometimes happens, especially in hot, enervating weather.

Once this demanding task is completed, the men and women work together to clear a path approximately two meters wide through the grasslands along the proposed fence line. The purpose of the path is not only to provide a clearing in which to erect the posts but also to create a firebreak, which is needed to protect the barbed wire and the pasture.

When the path is finished, men mark out a straight line by sighting along reed sticks (*Saccharum robustum*) placed every five meters to indicate the location of the post holes. After the men dig the holes, they thrust the posts into the ground, and where the corner posts are located they sometimes place stones in the holes to provide extra support. This activity is usually a male concern. Men state that their superior strength and ability are necessary to drive the posts securely into the ground. The women, however, are still occupied in distributing the fence posts along the cleared path.

By now, the intentions of those in the cattle project are clear to all. Previous exhortations demanding that the fence line be moved have obviously failed in some cases. Others look in dismay as their present gardens or past garden sites are enclosed. The fence line, which of necessity is as straight as possible for effective barbed wire tightening, gives little concession to traditional gardening patterns and occasionally bisects producing gardens. The problems of one elderly man are typical. He complained that the new cattle fence not only cut his existing garden in half but also enclosed many of his former garden sites that he hoped to re-use in the future. His protests were ignored. Some irate villagers also demand compensation for enclosure of their gardens, but this is to no avail. Occasionally, cattle bosses promise to put barbed wire around existing gardens to protect them from the cattle, but they never keep such promises. In more extreme cases, violence erupts. Almost invariably, *bisnis* wins out over subsistence.

The *fait accompli* is achieved when the posts are firmly in the ground. Then the men staple and strain the barbed wire, which must be fixed securely to the posts to prevent the cattle escaping. Women help occasionally during this phase by carrying rolls of barbed wire.

When fencing is completed, the project leaders inform the agricultural extension agents, who then come to inspect the completed product. The extension officer marks out a site for a stockyard and crush. The construction of a stockyard involves dragging extremely heavy posts from the forest, sometimes a distance of two kilometers. To lessen the burden, the people sometimes hire a vehicle from the extension officers to transport the large posts part of the distance. Both men and women work in dragging the posts.

In addition to providing labor, project members also contribute money to establish their enterprise. In the Kainantu District, the policy is to provide a PNGDB loan only after fences and stockyard have been built. Thus, project leaders and their followers must themselves purchase enough barbed wire to fence the boundary. The amount of barbed wire required varies, depending on the size of the enclosure, whether three or four strands of wire are used on the outside fence, and whether the project has a common border with another enterprise, thereby saving a part of the expense. The average cost of fencing a project was K420. In order to purchase wire, a cattle boss appeals to his followers to contribute money. The pooling of resources has precedence in the patterns of the traditional society in which individuals pooled their food and wealth to make prestations to other individuals and groups in competitive exchanges. Pooling has been widely employed in *bisnis* enterprises in Papua New Guinea to enable groups to participate in certain commercial ventures that would otherwise be beyond the financial capabilities of a single individual (Finney 1973).

Once the physical structure has been completed, the process of educating people in proper animal husbandry techniques follows. Extension officers provide a short two-week course in Kainantu and a longer two- or three-month course at the government livestock station at Baiyer River in the Western Highlands Province. Attendance at the courses from at least one project member is a prerequisite for obtaining a PNGDB loan. The cattle bosses themselves select the individual to be sent. Sometimes the project leaders attend, though they usually choose one of their young helpers because the bosses prefer to remain in the village to look after their interests, feeling that an absence of several months is too long. The problem with sending young helpers is their lack of influence, which makes it difficult for them to persuade other villagers to do the necessary project tasks when they return to Kapanara. At the courses, people receive both classroom lectures and field experience in pasture maintenance, fencing, proper veterinary care of cattle, stockyard construction, dehorning, castration, handling cattle, assisting in calving, and a host of other practical skills.

Another prerequisite for loan approval is to have two elders of the loan applicants' patrilineage cluster sign a "Clan Land Usage Agreement" acknowledging that their kin group has granted the loan applicants exclusive use of the land enclosed for the purpo of cattle raising. Even if disputes over the enclosed land remain, the elders give their formal acknowledgment in part because they previously consented when the bosses initially consulted them. In addition, many of those with complaints about the enclosures are from the other patrilineage clusters, and the elders are more favorably disposed toward their own cluster members.

After the Agreement is completed, the villagers can obtain a loan. The agricultural extension officers determine the amount of the loan to be received and the scheduling of the repayment period, and the cattle bosses sign as the applicants. For all seven cattle projects, the PNGDB approved the loan applications. The finance provided not only for the purchase of cattle, but also for pasture seed, veterinary supplies, and additional barbed wire for internal subdivisions (see Table 2.2). The interest on the loans ranged from 6 to 8 percent per year, with the earlier loans having the lower rates.

A few months after the project is established and the loan approved, livestock officers bring the cattle. First, they deliver the steers. The heifers and cows are brought several months later when project members have gained sufficient skill in handling cattle. The arrival of the steers marks a great public event, the official beginning of a large-scale *bisnis* enterprise. It is also the time to appease the ghosts of the ancestors, or *bana*.

The origin of the ritual dates back to 1969 when Bonafo's cattle arrived. Most of his cattle had been shipped from Australia and arrived in poor condition. Several died soon after coming to Ka-

TABLE 2.2

Amount of Loan per Project from PNGDB

Project Number	Amount of Loan (Kina)
1*	$\begin{cases} 1227 \\ 1200 \end{cases}$
2	3590
3	6010
4	4449
5	4320
6	3450
7	2798

* Received two loans.

panara. The villagers believed that either the *bana* were unfamiliar with these new beasts or the cattle had disturbed them, and consequently the *bana* either killed the cattle directly or were at least instrumental in their death. Occasional nefarious deeds of the *bana* are well known; for example, making loud noises near their graves can lead to serious illness. Supernatural forces are often associated with the untimely death of humans and pigs. Given the villagers' traditional epistemology, they were logical in linking the death of the cattle with supernatural forces because the cattle appeared to die for no apparent reason.[6] To appease the ancestors' ghosts, one villager devised a ceremony that appeared to be effective, because soon after he performed it no more of Bonafo's cattle died. Members of each of the seven village cattle projects subsequently performed the ritual to protect their cattle.

The ritual is performed soon after the steers arrive. First, a platform is prepared inside the stockyard, and some food and beer placed on it. The stockyard itself is decorated with flowers and leaves from the forest, and money is placed at various points inside the project boundaries. People blow bamboo flutes to notify the *bana* of the event. Then one of the elderly men who is familiar with the names of the ancestors shouts for all the *bana* to hear something like the following:

> Before, when you were here, you only had pigs. Now we have constructed a cattle project, and the cattle have arrived. We have prepared food for you, and you must come see this food, eat it, and take the money. You must be pleased with our effort. You cannot destroy or ruin our cattle. They must remain healthy. You must look after our cattle.

The people believe that if they do not perform the ritual and make the food offering, the cattle will become wild, people will not be able to herd them, and the *bana* might lead the animals into a ditch causing them injury.

After the speech is delivered, the feast begins. The leaders go to some effort to prepare enough food for all visitors, so that the latter can come and be pleased with the new project. The food offered to the ancestors is then eaten by the older men, usually those of the same patrilineage cluster as the cattle bosses. The old men actually demand that food be given to them at the cer-

---

[6] Since 1969, the villagers have learned more about cattle. During my fieldwork, they usually attributed the death of cattle to more secular causes, such as inadequate pasture or worm infestations.

emony. After all, they argue, they were the ones who were brave in wartime and defended the land against enemy attack. Now they are "giving" this land to the project members. Without the food presentation, the old men would feel insulted and harbor a grudge, which could possibly lead to unstated but undoubtedly unfavorable consequences for the project. The most adept sorcerers are the old men, but by making the food presentation, the leaders satisfy the elders. Food is also distributed to members of the other cattle projects and to other villagers attending the feast. When members of the last four cattle projects received their first shipment of steers, they all held their feasts on the same day. Not only was food distributed according to the principles noted above, but the agricultural extension officers were also invited. For the feasts that night, the people purchased about 25 cartons of beer and slaughtered 10 pigs. To kill so many pigs in one day is a rare event in Kapanara. It symbolized the great importance that the people attach to their *bisnis* enterprises.

Cattle enclosures are not static phenomena on the landscape. After a few years, project members sometimes increase the area enclosed, occasionally without seeking the permission of other villagers. They may decide to enclose more land either because the pasture in the paddock is insufficient to support their cattle or because they want to increase their prestige. The importance of the competitive spirit in relation to prestige is clearly illustrated by the following example. Referring to two newer projects that were larger than his, a cattle boss stated that he had decided to expand the area of his own project because he felt "*Nogut ol i winim mipela*," a phrase in *Tok Pisin* meaning "The others should not surpass our efforts." The phrase implies that "if the others do surpass our efforts, our prestige will be lowered."

MOTIVATIONS INVOLVED IN STARTING CATTLE PROJECTS

When I first visited Kapanara in 1976, almost everyone was enthusiastic about the cattle projects. Villagers were eager to become involved in this form of commodity production for several reasons. Cattle projects are one form of *bisnis*, and all the factors stimulating the Kapanarans' involvement in cash-earning activities are relevant to the adoption of cattle raising. Other motivating factors are peculiar to the cattle projects. Goals of the cattle project leaders, their helpers or followers, and the kin groups united as a whole must be distinguished, though motivations relevant to one of these may be applicable to the others.

Leaders enter the cattle business in part to earn an income. The only significant income-earning alternative within the village is coffee production, but prices received for parchment coffee were low until late 1975. One cattle project leader recalled that he started his project because

> I felt that the two cattle bosses were earning plenty of money. Bonafo too. Bonafo had butchered a steer and was marketing it. I went to see this, and he was making plenty of money. I myself saw this. So, I felt, if I had a cattle project I could sell cattle to people from other places who wanted to come buy and eat beef. I could have a source of income. One bag of coffee will sell for only a small amount of money. Sell just one steer, this will bring K200 or K300.

Another revealing event occurred in August 1976, when the same cattle project leader was giving a party for those who helped him in a previous fight that led to jail sentences for some of his friends. To provide food for the party, he told his brother to kill a steer. His brother was intoxicated, and finding difficulty in herding the cattle, he killed the first available animal, a cow. Upon learning of this action, the cattle project leader began to cry and then tore up some money and broke several arrows, both means of expressing sorrow. He lamented that to kill a steer is justifiable because a steer will not produce any more *bisnis* but a cow carries calves, and that means more *bisnis*. He felt that he lost the potential to earn K1,000 or K2,000 by the killing of his cow. Future income and *bisnis* potential died with the cow.

The monetary incentive, although very important, is not the only motivating factor. The leaders of a cattle project hope to gain prestige, which can be obtained in a variety of ways. They will, for example, limit the number of sales to ensure that at least a few animals remain so that their enterprise will not appear defunct, because, as a general rule, the more cattle in a project, the more impressive it is. Demonstrating the control of the flow of money in transactions is also important in attaining prestige. Traditionally, those at the focal point of substantial exchanges gained recognition and influence. In one day, three steers may possibly be sold for a total price varying from K600 to K900, a sum far exceeding the value of other transactions in the village. Though the cattle bosses may send a considerable portion of the proceeds from the sale to the PNGDB to repay the loan, they are still the focal point of the large transaction. In addition, having an entrepreneurial spirit, a willingness to participate in not one but several income-producing activities, demonstrates a strong commitment to *bisnis* and thus enhances one's prestige. Cattle projects are a significant form of *bisnis* in which to engage.

The followers or helpers expect to share in the profits of the enterprise. Also, project membership enables an individual to purchase a heifer or steer and place it in the paddock with the herd financed by the PNGDB. Many followers have done so. An initial investment of K100 can lead to a 100 or 200 percent profit within three years. Social considerations are also important. Groups involved in projects have meetings as well as occasional parties and feasts, and these are important social events. Not being a member deprives one of full participation in these social functions. This is not insignificant because the highlights of life for villagers are the public, social occasions.

Three other aspects appeal to leaders and followers alike. They all enjoy eating beef, and membership gives one the right to receive beef on certain occasions. In addition, strength and courage are key elements in the ideal man, and a villager has ample opportunity to demonstrate these characteristics when handling the partially tamed cattle, which are large and potentially dangerous. Finally, people often show concern for the future welfare of their children in stating that they want them to have income-producing activities and not to be worthless. Cattle project leaders and their helpers expect to pass on their respective rights in the projects to their children.

The motivations considered so far have pertained to individuals. Individuals combine to form social groups whose relationship with other similar units affects people's evaluation of cattle raising. The maintenance of group prestige in relation to other groups is a central concern. When one group starts a cattle project and another does not, prestige accrues to the former. In order to maintain equivalence in this atmosphere of competition, the second group must also start a cattle project.

Another major group concern in Kapanara is control over land. In the precontact era, boundaries separating the land of the various kin groups in the village were only roughly demarcated and were not important in daily affairs. Land was needed for subsistence gardening, but because land was plentiful, access to the means of production was not a problem. In the 1970s, people began perceiving that land had a new economic value as a cash-producing asset. Money can be obtained from growing perennial tree crops such as coffee, selling lumber, or selling the land itself to the government. The number of perceived potential monetary benefits that the land can yield are many, but the amount of land in the village is finite. Perceived scarcity inflates land values. Cattle boundary fences clearly demarcate sections of land, and the villagers feel that if one group does not erect a cattle fence on a particular site, another group might. Enclosing an area within a fence symbolizes a group's control over the land. For example, members of one patrilineage cluster were concerned that those of another were planting coffee on the former's land, so they erected a cattle fence to stop the spread of coffee plantings.

Ultimately, the members of Kapanara view themselves as a united group standing in contrast to other villages. At this level as well, the maintenance of prestige is an important consideration. To have outsiders come to Kapanara and see the area filled with *bisnis* enterprises such as cattle projects adds to the renown of the entire village. Kapanarans were particularly proud that their village had more cattle projects than most other villages in the Kainantu District.

I had expected that this general euphoria over cattle that I observed in the mid-1970s would endure. I was wrong. Enthusiasm had abated markedly by the early 1980s as the villagers became increasingly disillusioned with the consequences. As has happened in peasant communities throughout the Third World, increased commodity production in Kapanara disrupted traditional social relations of production, the local environment, and subsistence production.

1. Picking coffee is the major source of village income.

2. The drainage system reaches its highest level of complexity in valley bottom gardening (*aruka kai'a*). Men sometimes begin the arduous task of digging drainage ditches up to a year before planting.

3. Pigs wait to be fed outside their owner's house. A conscientious owner will feed her pigs once in the morning and again in late afternoon.

4. Villagers are frequently seen selling coffee to roadside buyers during the coffee flush while others watch.

5. Hungry pigs are a serious problem for the subsistence system. A gardener repairs his fence after pigs invade his plot and destroy many sweet potato mounds.

6. The task of planting sweet potato, the major staple, is performed largely by women.

7. View of Kapanara village. Tree-covered area in center is where the hamlets and coffee groves are located.

8. Man prepares to kill a pig for an exchange. The unsuspecting animals are eating raw sweet potato, their main ration.

9. Store-bought foods and beer purchased during a trip to Kainantu are consumed most frequently during the coffee flush.

10. Only men are involved in nailing barbed wire to project fences.

11. The first step in constructing a cattle project fence is cutting trees in a forest to prepare the fence posts.

12. Villagers use a traditional hammer and wedge to split sections of a tree into fence posts for a cattle project.

13. Women carry most of the heavy posts used to construct project fences. An average post weighs approximately 16 kilograms.

CHAPTER THREE

Cattle and Rural
Economic Differentiation

RURAL ECONOMIC DIFFERENTIATION is still in the early stages in Kapanara, not having reached the extent that it has in other Third World peasantries where it has resulted in the impoverishment and landlessness of certain segments of the community. However, focusing on an early stage of differentiation does provide a major advantage in being able to observe first-hand the processes at work. Government policy clearly plays a role, but stressing its importance does not require portraying villagers as inert, passive puppets dancing on strings pulled by relentless, inexorable, and exploitative external forces, as is characteristic of some studies of rural development. Rather, some Kapanarans were particularly adept in manipulating government policies and their relationships with government officials to foster their own economic interests at the expense of fellow villagers. In addition, concomitant with the expansion of commercial activity in the community, individualism, which both facilitates and reflects rural economic differentiation, was also increasing, being manifested in a variety of realms in village economic and social life.

Rural economic differentiation is best understood in the context of control over the production process. In particular, in focusing on changing social relations of production, we must consider those aspects between: "(a) man and means of production (especially land)—*property* relations; (b) man, his labour-power—*labour* relations, and (c) man and the product of his labour—basically the question of '*surplus*' [emphasis in original]" (Cliffe 1977: 205). To analyze changes associated with the introduction of smallholder cattle projects, we must examine differential access to and control over land, qualitative changes in labor relations, and the distribution of the surplus derived from the enterprises.

To understand alterations in patterns of economic differentiation, we must first consider the changing role of community leaders (big-men) in the social relations of production. Generalizations concerning leader-follower relations in Kapanara and the Tairora area are not applicable to the entire Highlands because there was and still is considerable diversity in the manifestation of the "big-man" political system. For example, in a few areas leadership was at least partly hereditary or ascribed, whereas in most others it was achieved through individual effort (cf. Standish 1978). In several regions, big-men had considerable control over other villagers' access to land and traditional valuables, but elsewhere where the system was more egalitarian they did not (see Feil 1982). Some researchers (e.g. Gerritsen 1979; Good 1979; Good and Donaldson 1980) argue that current economic differentiation associated with the commercial economy in the Highlands is a manifestation of the tendencies of the traditional big-man system in which leaders controlled other people's labor and access to land and valuables; for leaders to do so in the context of the resources available in the commercial economy as well is a logical outcome of traditional leader-follower relations characterized by inequalities. However, such an argument ignores significant variations in both the characteristics of the big-man system in different areas and in important historical changes. In Kapanara, relations between cattle bosses and their helpers are qualitatively different from those existing between big-men and their followers, and thus the current economic differentiation in the village cannot be viewed as a manifestation of the tendencies of the big-man system.

Village Leaders

In the precontact era, leadership in Kapanara as elsewhere in the Tairora region was achieved and based largely upon fighting prowess (Watson 1967). Success in competitive ceremonial exchanges did not play as significant a role in determining leadership as it did in the central and western Highlands regions. A big-man both persuaded and sometimes intimidated his followers, influencing them occasionally to provide resources for activities and exchanges that would enhance his prestige and power. People tolerated the machinations of a leader because he was their best guarantee of survival in a society in which intergroup warfare was endemic. Such a leader was the product of the masculine ethic in Tairora culture, which emphasized strength, forcefulness, and indifference to fear or threat (ibid.). Leaders, or big-men, were known as kempuka baiinti meaning "strong man" or "hard man."

At any one time, each of the three patrilineage clusters usually had at least one big-man.

In the precontact era, differences in the control of wealth did occur among villagers. Nevertheless, economic differentiation was not perpetuated from generation to generation for several reasons. A leader gave much of his wealth to others to satisfy old obligations or create new ones. Also, many of a leader's assets were distributed to a wide range of people upon his death, thus preventing differential accumulation of capital within particular families from passing to subsequent generations. In addition, a big-man's status was achieved and thus not necessarily inherited by one of his sons (cf. Standish 1978); unless an individual displayed the characteristics of a leader, he would not become a big-man.

With the establishment of colonial administration in the area in the late 1940s, warfare as a recurrent feature of intergroup relationships ceased, and ability in warfare could no longer mark a man for leadership, though courage and aggressiveness are still admired features of village leaders. Subsequent big-men have had less control over their followers.

The administration also created new positions of leadership. In the mid-1940s, administration officers appointed a *luluai* and an assistant called a *tultul* in Kapanara. As in other Highlands villages, their function was to help maintain order within the community and implement administration policies such as road building or constructing government rest houses. Each *luluai* subsequently appointed in Kapanara was also a traditional leader of one of the patrilineage clusters. The appointed village official was replaced by an elected one called the *kaunsil* or councillor in 1966 when the Tairora Local Government Council was formed. The purpose of local government was to bring *kaunsil*s from different villages together to discuss problems of local importance and to formulate policies on issues of common concern. Like the *luluai*s, the *kaunsil*s also helped implement administration policies within the village. The more recently established *Eria Komuniti* (Area Community), a local administrative unit composed of one or more villages, elects both a councillor to represent it at meetings of the Kainantu Local Government Council and a magistrate to settle disputes within the village. Although some leaders of the patrilineage clusters have been involved in such elected positions, other big-men have not. Men can achieve political and economic prominence within the village without the aid of elected office. Nor has appointed or elected office been a major influence on economic differentiation.[1]

When asked about current village leaders, Kapanarans list qualities that can be combined into five general headings: (1) the organizer and initiator of activities; (2) a persuasive orator; (3) the focal point of exchanges; (4) a generous man; and (5) protector of the cluster's land.[2]

A big-man is at the forefront of public activities. When a kinsman dies, the leader will tell his other relatives to prepare food for the mortuary feast and to pool resources to fulfill customary obligations to certain relatives of the deceased. If people are not planting enough gardens, he may urge them to rectify the situation. Before pacification, village leaders also initiated warfare activities. All these activities concern the welfare of the community. By initiating such activities, the big-man demonstrates his interest in the welfare of the people as well as his vitality and assertiveness.

To influence people to participate in activities requires competency in oratory. A big-man "talks strong"; by sheer force of personality he can motivate others. Ability in oratory is also essential for settling disputes, another concern of leaders.

The big-man is often the focal point of exchanges. Not only is he instrumental in pooling resources for prestations, but he is also responsible for distributing food and other items given to his kin group. When making such distributions, one or two other prominent members of the patrilineage cluster usually assist him.

Generosity was the least often mentioned characteristic of a big-man, but it is certainly significant. When kin groups are involved in exchanges, big-men are often the major contributors to prestations. In addition, they contribute generously to their followers' prestations, such as a bridewealth payment, though such gifts put an obligation on the recipient to reciprocate in some way in the future. People often visit the leader's house, knowing that food will be provided in plentiful amounts to all guests.

A big-man must have a sufficient supply of resources to maintain his ability to participate in exchanges that enhance his pres-

[1] Researchers in other areas of the Highlands have reported that appointed and elected officials have sometimes used their position for substantial economic gain (e.g. Brown 1963; Fitzpatrick 1980: 124).

[2] The role of the leader in relation to land is examined in the section on land tenure in this chapter.

tige and demonstrate his generosity. Having several wives enables a man to utilize more effectively his relatively extensive commercial holdings, prepare larger gardens, and raise more pigs. Part of the production from these activities is channelled into the exchange system. Two of the three leaders of the patrilineage clusters have three wives, and the third has four. These figures contrast with those for the village as a whole. Only 5 of 93 married men have three or more wives; of the two who are not cluster leaders, one is the son of a big-man and the other a cattle boss. Because young men either do not have enough wives, control enough productive assets, or have the relevant experience to become men of stature in village affairs, the prime of a man's political career is usually between the ages of 35 and 50.

In most instances, leaders are agnates of the patrilineage cluster, but if no agnate is suitable, others, such as the son of a female agnate, may become the leader. Sometimes, more than one big-man exists within a cluster, and as one leader becomes older and less dynamic, a younger leader emerges to take his place.

Big-men are supported by their followers who are usually agnates and, to a lesser extent, cognates and affines. These followers contribute labor and resources to the endeavors and prestations in which the leader is involved. Such activities bring prestige not only to the leader but to those who have helped him. Followers also occasionally assist the big-man in his garden and do errands for him. To reciprocate, the leader contributes to his helpers' exchanges, grants them temporary use rights in his gardens, provides food and sometimes beer for them when they visit him, and occasionally agists female piglets with them with the understanding that the followers may keep some of the future progeny. A big-man's support is dependent on not becoming "over-extractive" (Sahlins 1963); he must not demand too much from his followers and must reciprocate adequately for the help he has received.[3] The big-man cannot command or force his followers to help. If over-extraction continues to characterize the relationship between leader and follower, the latter can withdraw the support on which the current big-man depends.

In most daily affairs, individual households are left to their own initiative to determine their commitments and activities. Thus,

[3] Although this generalization is applicable to Kapanara today, it is not completely valid for the precontact era. As Watson (1967) demonstrated, a great warrior who had a tendency to be over-extractive was tolerated.

big-men attempt to assert their influence on only a limited number of occasions. Current leaders in Kapanara lack the power characteristic of big-men in many other Highlands societies (see Standish 1978), perhaps because of the smaller size of the political units here and the absence of large-scale ceremonial exchanges that are found elsewhere.

The status of a big-man is achieved, though it would be inaccurate to claim that all men start out equal in the quest for leadership. Having a father who purchases several wives for his son and leaves many *Pandanus* nut trees, *Areca* palm trees, and coffee plots gives some men an initial advantage. However, unless an individual displays the characteristics of a big-man, he will not become a leader.

The relationship between big-men and economic differentiation has changed since the precontact era. Although a leader today has less control over his followers' labor and resources than his precontact counterpart, his own wealth is more durable in the form of cash and coffee trees. However, preserving a current leader's wealth for his children is still somewhat circumscribed. As in the precontact era, much of his wealth is given to a wide range of kinsmen upon his death, and that given to his children is divided among his sons and, to a lesser extent, his daughters.

Cattle Projects, Leadership, and Labor

Kapanarans recognize two types of leaders. One is the leader of the patrilineage cluster whose characteristics have been described. The other is the cattle boss. Big-men function within the context of interpersonal and intergroup relationships defined partly by patterns of kinship, descent, and exchange, whereas the cattle bosses are engaged in managing a business enterprise. This distinction between the domains, however, is somewhat spurious. Three big-men are also cattle bosses, and kinship and descent are important principles in recruiting followers to the cattle projects. Thus, certain aspects of leader-follower relationships based on the big-man model might be expected to occur in the organization of the cattle projects.

Nevertheless, the additional factor of a specific development package accompanying the cattle projects presents the basis for changing the nature of leader-follower relationships. Social relations of production within the village are heavily influenced by the links with and policies of government agencies—specifically, the PNGDB from which people obtain credit to purchase the cattle and the agricultural extension service whose officers act as the Bank's field agents and supervise the cattle projects. In addition, a relatively large-scale economic enterprise, the cattle project, has to be started and maintained over several years, and this requires organizing personnel to provide labor and finance. Given the relatively large scale of the cattle projects compared to traditional economic activities, one man alone cannot start and operate a cattle project because of limited financial resources, land, and time; he must call upon the help of others. As Langness (1975: 83) noted for a cattle project near Goroka:

> But appearances to the contrary, there is no precedent for a clan-wide, or even a sub-clan-wide common project of this kind. Houses, gardens, pigs and even coffee gardens are all owned by individuals. While people do co-operate to help each other, they do not have communal projects.

Although people in Kapanara previously cooperated in several communal ventures, such as goldmining and operating PMVs and trade stores, the nature and scope of the organization of the cattle projects introduced elements without precedent. Finally, the potential annual revenues from a project far exceed the average household annual income. Qualitatively new relationships have emerged between cattle bosses and their helpers.

The distinction between cattle boss and follower is an emic one. In *Tok Pisin*, the boss is called the *papa bilong banis* ("father of the fence or project"), *fama* ("farmer"), or more rarely *masta* ("master"). The term *fama* is used only to refer to the bosses and not horticulturalists in general; extension officers use the term widely, and thus it is not peculiar to the village. *Masta* is generally employed to refer to any European or employer; it has always implied relations of superordination and subordination. Extension agents, cattle bosses, as well as their followers use the term *boi* to refer to project helpers. *Boi* was also the term of address once used by Europeans to refer to nationals and connotes an inferior status.

Cattle bosses also adopt behavioral characteristics that distinguish them from their followers. Most project leaders do not eat beef from cattle in their project, declaring that the bovines are like their children; they have looked after them and fed them and thus cannot eat them. For similar reasons, bosses usually delegate to their followers the task of killing their cattle. Occasionally, just before an animal is killed, a boss ties money to its horn or rubs money in the bovine's nose as a gesture of sorrow and personal loss; the person who kills the animal usually keeps the money. One cattle project leader clearly demonstrated his sorrow over the unexpected death of one of his beasts by rubbing colored clay on his body, a traditional form of mourning. These actions clearly differentiate cattle boss from follower.

The cattle bosses are the project managers. Like a big-man, a cattle boss is at the center of activity in initiating work on the projects. He is responsible for making sure that the loan is repaid, and he occasionally seeks the aid of his followers to complete the tasks necessary to maintain the projects: clearing regrowth under the fence line every few years to create a fire break and to ensure that the wooden posts do not rot too rapidly, tightening loose barbed wire, constructing additional internal subdivisions for rotational grazing or to facilitate the yarding of cattle, planting improved pasture, replacing rotten fence posts, and rounding up cattle on various occasions. Although the potential tasks are numerous, most people work on average only one or two hours or less per week on the projects after the arduous task of fence building has been completed. In most cases, followers will not work unless the boss informs them of the necessary tasks. For the parties that project members stage on various occasions, such as at Christmas, the bosses tell their followers to pool money and prepare food.

Followers are motivated to work in the projects for several reasons. Obligations arising from kinship and affinal links are influential. People also want to maintain the prestige of their group. However, the expectation to benefit materially from project revenues is their most important consideration. A possible alternative explanation is that a big-man, in contrast to other villagers, is a powerful leader of the community and can thus obtain free or inexpensive labor from his followers. This has happened elsewhere in the Highlands (see Finney 1973: 63), but such an explanation is not valid for projects in Kapanara. First, patrilineage cluster leaders are involved in only three of the seven projects. The other project bosses are entrepreneurs and respected members of minimal patrilineage units, and many of them are able to mobilize just as much labor as the big-men can. Second, when big-men request that their followers provide help in their subsistence activities, they reciprocate by providing a small feast of store-bought foods and beer. The cash amount expended by the

big-men for the feast compared with the labor input of their followers is sometimes greater in value than agricultural rural wages
in the area (K10-K22 per week per adult in 1977). Thus, big-men
of the clusters do not generally obtain free labor for long periods
of time. Reciprocation is expected.

By expecting to benefit from supporting their cattle bosses, the
followers thus imitate aspects of traditional leader-follower relations. However, the problem of becoming over-extractive and
losing the support of one's followers is less serious for cattle bosses
than for big-men for a variety of reasons. The position of the cattle
bosses is strengthened by the nature of the links with government
agencies, a relationship that the bosses manipulate skillfully. The
followers have a substantial investment of time, money, and land
in their project, thus making it undesirable and difficult for them
simply to leave one project and join another when dissatisfied
with their boss. In addition, once a project is established, a cattle
boss is less dependent on his followers than a big-man because
he can use project resources and revenues to hire other laborers.

Labor relations are influenced by the policies of the PNGDB
and the agricultural extension officers. On the PNGDB loan application form there is space for up to three loan applicants, though
in most cases in Kapanara the two cattle bosses signed as applicants. The names of these two loan applicants are also placed on
the Clan Land Usage Agreement, which government officials read
before the villagers (see Fitzpatrick 1980: 115). Almost all subsequent official contacts from the PNGDB and the agricultural
officers to the smallholder projects are made to the loan applicants. When various forms have to be signed, the agricultural
officers contact the cattle bosses. When villagers occasionally sell
project cattle to the abattoir in Goroka or to the government
livestock station in Kainantu, the government sends money from
the sale to the cattle boss.[4] Loan statements from the PNGDB
are sent to the cattle bosses, and in most cases, only the cattle

[4] When villagers send cattle to the abattoir in Goroka, the abattoir sends the
proceeds of the sale directly to the PNGDB which, in turn, refunds a small portion
of the sale to the cattle bosses. The PNGDB retains the rest of the money as a
loan repayment. To facilitate the marketing of cattle in the Kainantu area, extension officers sometimes have obtained cattle from smallholders to sell to other
villagers who have expressed an interest in purchasing cattle. They keep the cattle
temporarily in the livestock station near Kainantu before selling them. The proceeds of the sale are usually given to the cattle bosses, though at other times
extension officers have sent the money directly to the PNGDB without asking
the cattle bosses' permission.

bosses attend meetings of the Kainantu Bulumakau Association. Thus, a whole range of external contacts is focused only on the cattle bosses, a pattern typical of most smallholder cattle projects in the country (McKillop 1976a). The focusing of extension and PNGDB inputs on a limited number of individuals is influenced by the government's desire to streamline its work effort and have specific individuals who are responsible for loan repayments. As a result, many other villagers believe that the project leaders are officially recognized by the PNGDB and the agricultural extension officers, and thus their role as cattle bosses is legitimated and their influence is enhanced.

The cattle bosses believe that their links with the agricultural extension agents will ensure the support and backing of the government if conflicts in relation to the projects develop with their followers or with other villagers. This understanding is related to a rather vague notion held by the villagers that recognition by, familiarity with, and access to government agencies are important in obtaining influence and support from them. Because government agencies are felt to have ultimate authority in determining the outcome not only of certain disputes related to the cattle projects but of a whole range of potential problems, such as adultery, murder, and abuse of elected office, their recognition and support are highly valued. Both cattle bosses and other villagers request aid from agricultural extension officers to settle arguments concerning the cattle enterprises. In most cases, these government officers tell both the cattle bosses and the other villagers to settle their disputes in the village court, because the officers do not want to become involved in village affairs. However, sometimes the extension agents either give support or what is interpreted as support to the cattle bosses, thus reinforcing the belief in the village that the project leaders have a privileged relationship with the extension agents. Cattle bosses occasionally remind their followers of their formal relationship with the government. During a dispute, for example, one boss told his followers: "Only I put the [PNGDB loan] contract with the agricultural officers. You do not have a name too [did not sign the contract]."

The cattle bosses use their links with the government to their own advantage to mobilize the labor of their followers. If the leaders are having difficulty in obtaining labor, they sometimes inform their helpers that an agricultural officer has told them that they and their followers must all work on a particular task designated by the extension agent, such as fixing the stockyard. At other times, bosses remind their followers that an officer has told them that if their followers do not help with the cattle project, the followers will lose their membership in the project, forfeit any potential profits to which they would normally be entitled, and have their cattle removed from the project. The followers cannot readily verify their project leaders' statements. The cattle bosses' justification of actions and requests by referring to directives from agricultural extension officers is partially effective because it is patterned on the previous form of relationship with the government; patrol officers have told villagers what to do for so long that orders from the government have become part of the perceived routine of interaction.[5] Although the cattle bosses do not always obtain compliance from their followers when so justifying their actions, their relationship with the government does give them added leverage in convincing villagers and mobilizing labor.

Cattle bosses do not always have to threaten or cajole their followers into providing labor. If the helpers believe that their bosses have been fairly equitable in distributing the profits of the enterprise and have not been too demanding in their requests for labor and finance (i.e., have not become over-extractive), the helpers willingly provide labor and finance. Bosses can obtain continued support from their followers in a variety of ways. Instead of marketing a steer or cow that has died unexpectedly, they can let their followers consume it without charge. Instead of sending all the proceeds from a cattle sale to the PNGDB, they can distribute some of the money to their helpers. A boss can also subsidize a follower's purchase of a calf from the project herd by reducing the sale price by K10 or K20.

Other actions by the cattle project leaders are designed not to provide an adequate return for their followers, but to gain the maximum benefit for themselves. In other words, the phase of over-extraction is reached when the bosses do not adequately share the project revenues with their followers, demand too much labor without reciprocating, or in their followers' view misap-

---

[5] This pattern of interaction with the administration was widespread in the Highlands. For example, in discussing rural commercial activity in the Southern Highlands, Harris (1974: 3) observed that "information is both passed on by Administration officials and accepted by residents in a manner akin to orders rather than advice and the Huli seem content for this style of management to continue."

propriate project revenues for their own personal use. At such times, the followers cease providing labor.

As an alternative to ceasing work temporarily on the projects, the followers can withdraw from one project and attempt to join another in the hope of obtaining better returns. Only two men in Kapanara have done so. They responded to their cattle bosses' failure to distribute proceeds from the project. Followers rarely renounce completely their ties to a particular project because many have substantial investments in the project and they would lose several benefits. Villagers believe that when the loan is repaid a project member should receive a cash amount equivalent to his initial investment plus a sizeable, though unstated, additional amount.[6] Occasionally, followers receive free beef when cattle from their project die unexpectedly. Followers can purchase their own animals and place them within the project boundary fence. Withdrawing from the project may mean receiving back only the initial amount of money invested; all else might be lost. Indeed, if the relationship between a boss and his follower is hostile, the latter may receive nothing at all. In addition, it is difficult to join another project after the fences have been erected. The followers of established projects do not want others to join because they have already finished the most demanding task themselves and the addition of new members would reduce the potential returns to each helper.

Whereas followers are dependent on the state of leader-follower relations for their returns, the bosses are less dependent on their followers once the project is constructed. With his control of most stock and project revenues, a cattle boss has enough resources at his command to hire temporary labor to perform tasks on the project. One cattle boss almost exclusively uses such labor, repaying them with either cash, beer, or beef; most of his previous followers have either left the project or rarely work because of his failure to reciprocate for their work. A few other project leaders in Kapanara also on occasion hire temporary labor. Similar labor patterns occur in other projects in the Eastern Highlands Province as well.

[6] In other enterprises in which people have pooled resources such as in trade stores, some managers have returned a 100 percent profit on the initial investment, though others have returned only the original amount received. Thus, no standard procedure for repaying investments exists, though the followers in the cattle projects expect a considerable return because they have contributed so much labor and land.

Cattle Projects and the Distribution of Wealth

The potential for accumulation and differentiation in the village is heavily influenced by the links with and policies of government agencies—specifically the PNGDB and the agricultural extension service. The Bank preferred to lend money to projects having the capacity to hold 10 to 15 breeders, though smaller projects in Papua New Guinea also received funding. A variety of factors influenced the Bank to emphasize relatively large smallholder projects. The International Development Agency, an affiliate of the World Bank, provided much of the credit to the PNGDB to finance smallholder cattle projects; it originally stressed the importance of lending to projects with a 15-breeder capacity to ensure that they would be large enough to be commercially viable. Because extension agents have to complete a complex of forms for each loan application, they encourage larger projects.

> More importantly, the complexity of the credit package oriented extension agents toward large projects which would provide a better return for the effort involved. Unofficially, Development Bank administrators concede that loan applications for less than $500 are too costly to administer. . . . (McKillop 1976a: 11-12)

Because of the relatively large size of the enterprises, the potential annual revenues of a project far exceed the average Kapanaran household income. In a large project, for example, five to eight progeny can be produced each year, some of which could be sold eventually for K300 per head.

The bosses do not exclusively own the cattle projects, though as initiators of the enterprise their rights are superior to those of their followers. The followers, who have contributed labor and money to establish and maintain the project and help repay the loan, also have an interest in the enterprise. For example, followers in three projects brought complaints against their cattle bosses to the village court, alleging that their bosses used large amounts of project revenues for their own personal use. The desire of the followers to ensure that their interests are protected is the rationale for such actions; the profits from the project do not belong exclusively to the bosses.

Indeed, the expectation to benefit materially from the proceeds of the project is the main reason followers initially provide labor. Villagers view cattle as a very significant potential source of income, because prices for cattle exceed amounts normally obtained selling other commodities. Sale prices for cattle vary considerably. In 1977, project leaders sold weaners for K100 to K150 per head depending on size. Mature steers were sold for K200 to K300 each. People from Kapanara wishing to invest in cattle purchased most of the weaners. Most of the steers were sold to people from other villages, to the government-owned abattoir, and to project members wishing to consume beef. Cattle are not cut up and marketed in pieces unless an animal dies unexpectedly or is close to death; prices received for marketing a large animal in such cases range from only K20 to K80.

Most of the income from the projects has been sent to the PNGDB to repay the loans (which are still outstanding) and has not been divided among the cattle project leaders and followers. Nevertheless, the initial distribution of the financial benefits of the enterprises is revealing and can be used as a likely indication of future trends.

Most followers expect to receive the major return on their investments of labor and cash after the loans are repaid. Whether their returns will be in one lump sum or in a series of payments for the duration of the enterprise has not yet been decided, but a majority expect to receive cash as opposed to cattle. Villagers agree that after a project loan is repaid, cattle purchased with funds from the PNGDB and the progeny of these cattle will belong to the cattle bosses.

An accounting of the financial inputs and returns is made in Table 3.1. A few estimates had to be made, but the results are very close approximations. For comparative and explanatory purposes, data for each of the seven projects are included separately. The data are aggregated into two groups: cattle bosses and followers. Individual variation within each project is not considered. Each enterprise has two cattle bosses, except number 2, which has one. Finally, most but not all followers have contributed financially to the building of the projects and loan repayments.

In the three oldest projects, the bosses as a group contributed more money than the followers as a group to establish the project (columns 2 and 3).[7] The situation is reversed in the four most

[7] Data were obtained by interviewing, surveying the project boundary fences, and cross-checking the data. The followers were asked how much they contributed and then, as a cross-check, the bosses were asked how much each follower contributed. Only a few major discrepancies were found and were further cross-checked against other interviews. The most unreliable estimates were given by

TABLE 3.1

Financial Data concerning the Cattle Projects as of November 1977 (in Kina)

	(1)	(2)	(3)	(4)	(5)	(6)
			Initial Contributions			
Project Number	Years Stocked	Bosses	Followers	Followers' Households Contributing	PNGDB Loan	Total Gross Income
1	8	240	30	1	2427	2383
2	4	330	40	3	3590	2525
3	4	404	146	7	6010	4289
4	2	200	450	28	4449	1335
5	2	180	235	7	4320	1354
6	2	284	436	8	3450	1000
7	2	150	200	17	2798	464

(7)	(8)	(9)	(10)	(11)	(12)
		Distribution of Proceeds			
Loan Repayments Plus Incidental Expenses[a]	Loan Repayments by Followers	Revenue Given to Followers	Revenue Kept by Bosses[b]	Net Income of Bosses[c]	Net Income of Followers[d]
2323	230	0	290	50	−260
1850	68	20	723	393	−88
3180	195	155	1149	745	−186
1510	90	16	−101	−301	−524
1280	0	14	60	−120	−221
930	0	90	−20	−304	−346
280	100	0	284	134	−300

[a] 'Incidental expenses' include costs for various services and supplies not covered by the loan. For the first three projects, K200 has been added for such expenses and K50 for the last four projects. These figures are approximations.

[b] Column 10 = (Columns 6 + 8) − (Columns 7 + 9).

[c] Column 11 = Column 10 − Column 2.

[d] Column 12 = Column 9 − (Columns 3 + 8).

Source: Grossman 1983.

recently built projects, which are listed last in the table. However, the cattle bosses in the four newest projects justify their status in relation to their helpers by stating that they have contributed more money and labor than any individual helper. In most cases, this is true *at an individual level*, but not at the aggregate level. Except in two cases, no individual follower can assert that he contributed more money than his cattle boss, though the followers as a group could make such a claim against the bosses as a group in the last four projects but never did so (see Figure 3.1). The advantage of having a large number of followers contributing small amounts of money is readily apparent.

The followers have also made significant contributions from their own financial resources in helping to repay the loans (Table 3.1, column 8). Project leaders and their followers usually pool money to make the first few loan repayments, and afterwards revenues from the sale of cattle are used to repay the loan. No data are listed for the bosses because it is impossible to discover whether they used their own income from other sources or that derived from the projects to contribute to loan repayments.

Cattle bosses have not distributed much project income to their followers (column 9). Although the latter do not expect their major return until after the loan is repaid, they have been clearly dissatisfied with the small amounts given to them so far, as is evidenced by their numerous complaints. The cattle bosses have kept for themselves most of the proceeds not sent to the Bank (column 10). In one project, the two leaders retained approximately K1,149. In the two cases where a minus sign appears in column 10, the bosses have spent more money on repaying the loan than they retained for themselves. Their financial situation will improve as more cattle are sold. Aggregated data from the first three projects, which have existed the longest, are instructive; the bosses have contributed 58 percent of the money to start the projects and repay the loans, whereas they have kept for themselves 93 percent of the revenues not sent to the PNGDB.

Also revealing are data concerning the net incomes (or deficits),

---

the bosses on their own contributions. However, because over 90 percent of the money to start a project is used to purchase barbed wire, it was possible to cross-check the bosses' statements. All the projects were surveyed with compass and tape, and the number of wires (three or four) on each length of survey was recorded, thus yielding the total amount of wire purchased. By knowing the cost of the wire at the time of purchase and the contributions of the followers, it was possible to estimate the contributions of the bosses themselves.

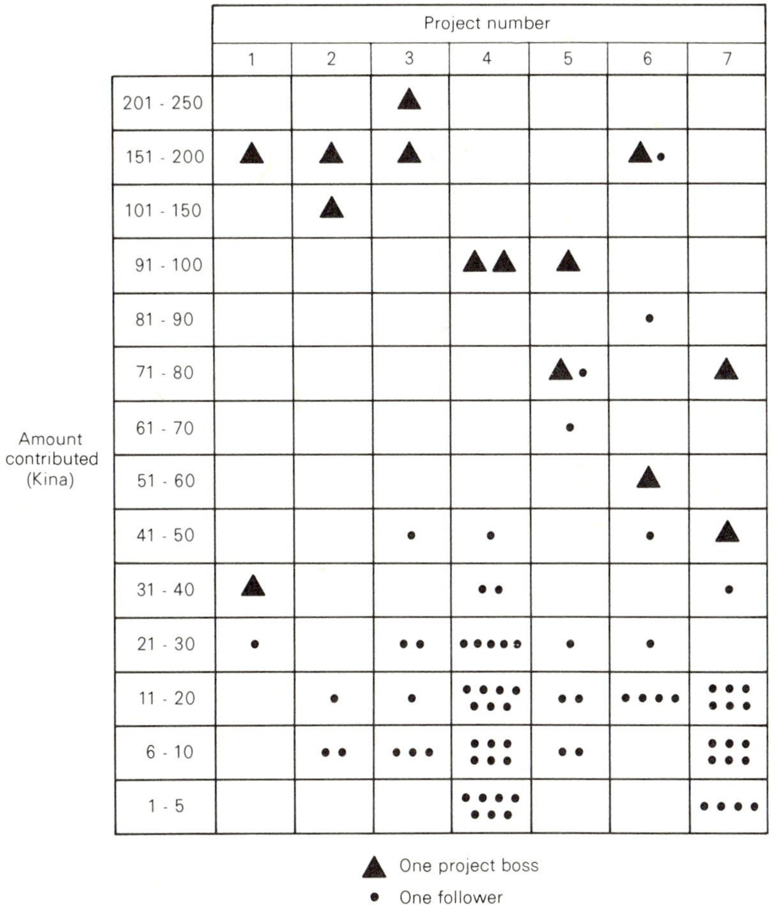

Amount contributed (Kina)	Project number						
	1	2	3	4	5	6	7
201 - 250			▲				
151 - 200	▲	▲	▲			▲ •	
101 - 150		▲					
91 - 100				▲▲	▲		
81 - 90						•	
71 - 80					▲ •		▲
61 - 70					•		
51 - 60						▲	
41 - 50			•	•		•	▲
31 - 40	▲			• •			•
21 - 30	•		• •	• • • • •	•	•	
11 - 20		•	•	• • • • • • •	• •	• • • •	• • • • • •
6 - 10		• •	• • •	• • • • • •	• •		• • • • • •
1 - 5				• • • • • • •			• • • •

▲ One project boss
• One follower

Figure 3.1. Frequency Distribution of Contributions by Individuals to Start Cattle Projects

defined as the total financial inputs compared with the cash returns, of the bosses and followers respectively (columns 11 and 12). The bosses again have obviously fared much better than their followers. In three of the seven projects, the bosses have showed a net gain, whereas in every project the followers as a group have not received as much money as they had invested. Even when both the bosses and followers have deficits, the deficits of the bosses as a group are less.

Although eventually the followers in some projects may possibly receive more money than they have invested, as a group they will never receive as much as their bosses will. That is, within each project, the combined net income of the cattle bosses will always be greater than the combined net income of all the followers. Clearly, the cattle project leaders are and will continue to be the chief beneficiaries of the K500 to K2,000 that each project has the potential of generating annually from the sale of its herd— a rather large amount in a village in which the per capita income in 1977 was approximately K75. The potential for the emergence of greater inequalities in monetary wealth than previously existed is evident.

The loan-repayment process itself enables cattle bosses to accumulate financial wealth. According to disgruntled followers, cattle bosses claim that they cannot distribute much revenue from the projects because of the need to repay the loans. However, because of the exclusive nature of contact between cattle bosses and the outside world, the followers are not sure whether cattle bosses actually make loan repayments or keep the money for themselves. Indeed, followers' complaints that cattle bosses have dishonestly kept project revenues intended for loan repayments are common throughout the Highlands.

To complete the accounting of the benefits that the followers receive from the enterprises, two additional advantages not directly related to the income potential of the general herd controlled by the cattle bosses must be considered. One is free beef. The other is the profit obtainable from individually owned cattle.

Cattle bosses have given their followers free meat or entrails on only a limited number of occasions, considering how long the projects have been in operation (Table 3.2, column 2). Only once did a project boss specifically kill a steer to reward his followers for their work efforts. On all other occasions, the helpers received free meat or entrails only when an animal died unexpectedly. Sometimes the bosses gave them only the pieces of beef that were not sold, though at other times the helpers received the entire animal to consume. In one project, the followers received free meat 13 times, but only because the project was heavily over-stocked and the breeder mortality rate was very high.

The major benefit that the project members have received so far is the ability to purchase their own animals and run them with the rest of the project herd. As of November 1977, 21 followers owned their own cattle; only two helpers owned more than one animal (column 3). Some purchased a steer, expecting a return on investment in approximately 24 to 30 months. Three

TABLE 3.2

Benefits Other Than Money Received by Followers
as of November 1977

Project Number	(1) Total Number of Cattle	(2) Number of Occasions Beef or Entrails Received Free	(3) Number of Followers Who Currently Own Cattle
1	26	1	1[b]
2	20	5	1
3	44	13	6
4	22[a]	2	7[b]
5	25[a]	1	3
6	16[a]	0	3
7	9[a]	0	0

[a] Project is not fully stocked. More cattle funded by PNGDB loan will be received.

[b] One follower owns two head of cattle.

helpers each bought a steer for K100 in 1975 and sold them at an average price of K287 in 1977. After repaying the men who lent money to help purchase the steers, these followers kept the rest of the proceeds. Others preferred to purchase heifers; the wait for a monetary return is longer, but the potential profits are greater when the progeny are eventually sold. Cows have a much higher mortality rate than steers[8] because of lactation stress (Leche 1977: 11), and thus an investment in a heifer is more risky. The benefits that the followers will receive from the sale of their own animals, however, will be small in comparison with the proceeds that the bosses will obtain from the sale of their own herds.

The accounting so far has only been concerned with the returns on money invested. If the labor inputs into building a project and the subsequent maintenance are also considered, the comparative returns to the followers are even less promising. It is impossible to reconstruct how much time each villager spent in the past working on the projects, but diaries were kept during fieldwork

[8] I estimate that on smallholder cattle projects in the Highlands, the breeder mortality rate is 15 percent and that of steers is 3 percent.

concerning the number of times individuals worked in two projects. The results (see Figure 3.2), which are aggregated into households, parallel those in Figure 3.1. As a group, the followers' households contributed labor more frequently than the bosses' did, but except in one case, the bosses' households worked more often than did any one of their followers'. Such differences only accentuate the disparity in the potential returns between the followers and the bosses. This pattern of labor inputs, of course, would not be applicable to projects in which most or all of the helpers no longer provide assistance, though it most likely would have been characteristic in the earlier stages of such projects.

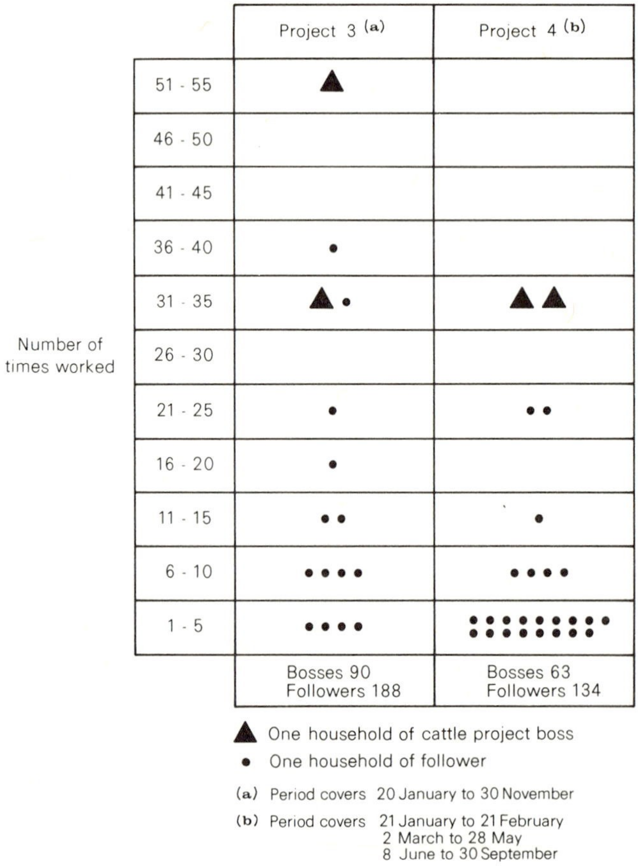

		Project 3 (a)	Project 4 (b)
	51 - 55	▲	
	46 - 50		
	41 - 45		
	36 - 40	•	
	31 - 35	▲ •	▲ ▲
Number of times worked	26 - 30		
	21 - 25	•	• •
	16 - 20	•	
	11 - 15	• •	•
	6 - 10	• • • •	• • • •
	1 - 5	• • • •	• • • • • • • • • • • • • • • • •
		Bosses 90 Followers 188	Bosses 63 Followers 134

▲ One household of cattle project boss

• One household of follower

(a) Period covers 20 January to 30 November

(b) Period covers 21 January to 21 February
2 March to 28 May
8 June to 30 September

Figure 3.2. Frequency Distribution of Number of Times Households Worked on Cattle Projects. *Source:* Grossman 1983.

The cattle bosses have another major advantage—the exclusive ability to give away in ceremonial exchanges both the cattle originally purchased with funds obtained from the PNGDB as well as the progeny of these cattle. This privilege not only gives them an additional source of income but enhances their prestige as well. Generosity in exchanges is highly respected, and giving a steer in an exchange is a mark of generosity. In several instances, a cattle boss killed a steer to provide a lavish feast on a special occasion and distributed the meat to those who attended the party. More frequently, however, project leaders killed and gave a steer in a ceremonial exchange with the expectation that money would be reciprocated by the recipients. The amount returned was usually between K250 and K300. These exchanges are patterned on traditional prestations involving pigs, except that now money and cattle are involved. Followers feel that their bosses should use most of the money from the exchanges to help repay the loan, but this does not always happen.

Cattle bosses are also able to sponsor an event, called a *singsing* in *Tok Pisin*, in which one or more groups are invited to come in traditional dress and dance within a specially built round enclosure. The *singsing* has existed since the precontact era, but the custom of charging participants a cash entrance fee started only in the 1970s. Depending upon the personal and kinship links to the sponsor of the *singsing*, guests are charged up to K50 each. Guests dance around a large bonfire from night until dawn and are given food and beer throughout the night; when ready to leave in the morning, they also receive additional beef, pork, and other foods. Prestige accrues to the sponsor if he and his helpers provide much food and beer for the fee-paying entrants, and killing a steer for the feast enhances the occasion. Because of the expense of providing much food and beer, some *singsings* are not very profitable financially, though others clearly are, with project leaders earning a net return (taking into consideration the cost of purchasing food, pigs, cattle, and beer) of K200 or K300 or more. One or two *singsings* are usually staged in Kapanara each year; nine of the ten held from 1974 to 1981 were sponsored by cattle bosses. Most other people do not have either the accessibility to cattle to use or the ability to influence the many others whose help is needed to obtain food and firewood and to construct the *singsing* enclosure. With his followers in the project as the core of his support, a project leader is more able to stage such a festivity, which brings him renown, prestige, and additional cash.

PROPERTY RELATIONS

In addition to obtaining a disproportionate amount of income from the cattle projects, the bosses have also enhanced their position in relation to land, the crucial means of production. They now assert more control over land within the fenced paddocks than big-men do with their cluster territory.

To understand qualitative changes in control over the means of production requires an analysis of the traditional land tenure system. However, the delineation of traditional rules of land tenure does present certain difficulties. Even the term "rule" itself is misleading because it implies predetermined procedures and principles to be applied to all subsequent cases. As Harding (1972: 606) observed for Papua New Guinea communities:

> Since principles of land tenure are not codified, and since in disputes there was formerly no agency independent of and superior to the litigants themselves that was charged with preserving and interpreting principles, customary rules are subject, to a marked degree, to the pressure of circumstances and dominant interests.

Description of the system is also difficult because it is in a state of flux due to changing economic and social conditions, and different individuals and groups have vested interests in perpetuating different viewpoints. As a result of these problems, the following characterization, of necessity, imparts more orderliness to the system than exists in reality.

Each patrilineage cluster is associated with a particular tract of land within the village territory where its ancestors first settled when they entered the valley (see Map 3.1). Each cluster's land is further divided into strips associated with a patrilineage or group of patrilineages. The strips run roughly parallel to each other from the forest to the grassland so that all groups control land in both environmental zones. These divisions do not imply exclusive control because members of other patrilineages and patrilineage clusters also have rights in the strips.

A consideration of the antecedent, historical conditions that have influenced land use and settlement will aid in understanding land tenure. Before pacification, group survival was the major concern in response to the constant threat of warfare. Members of the three patrilineage clusters often lived near each other in palisaded hamlets. They also planted many gardens in close prox-

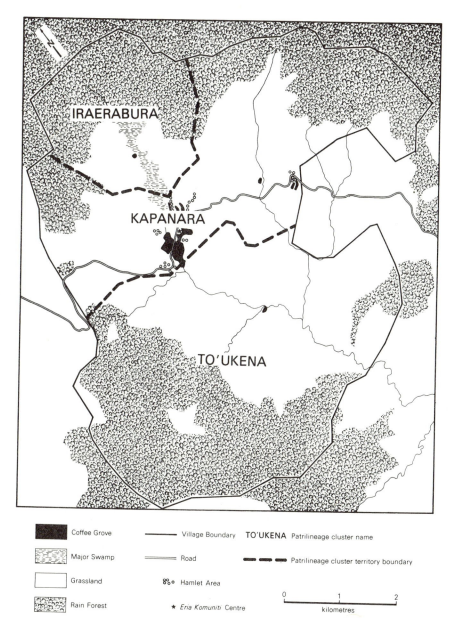

IRAERABURA

KAPANARA

TO'UKENA

	Coffee Grove	——— Village Boundary	**TO'UKENA** Patrilineage cluster name
	Major Swamp	═══ Road	▬ ▬ ▬ Patrilineage cluster territory boundary
	Grassland	8°₀ Hamlet Area	
	Rain Forest	★ *Eria Komuniti* Centre	0 1 2 kilometres

Map 3.1. Patrilineage Cluster Territories

imity to the hamlets because such areas were less susceptible to surprise enemy attack and people could more readily seek shelter within the hamlets. Security in numbers was the prime consideration, and given the abundance of land, the people of one cluster did not prevent members of another cluster from gardening on their land. In addition, villagers often changed hamlet sites because of defeat in warfare, and thus members of different patrilineage clusters lived together in several different areas of the village territory. The land immediately adjacent to sites where large aggregations from different clusters lived together in the past is today considered as not being the domain of any one cluster, but the property of all Kapanarans because land rights established by cultivation in these areas are so mixed. Thus, although the village territory is divided generally into three tracts of land, each associated with a patrilineage cluster, a few previous settlement sites within the territory of the clusters are not considered as part of the domain of any one cluster.

The big-man of each patrilineage cluster acts as the guardian of the cluster's territory. In this role, the leader is called the *bata'ara raa'aikibae baiintira* or the "man who supervises land." He does not direct land-use activities or allocate land, but instead he performs more generalized functions. The leader protects the cluster's territory against the claims of members of other clusters as well as other villages; thus, when one big-man saw people of another cluster attempting to enclose his cluster's land within their cattle project, he vociferously objected and they changed the direction of the boundary fence. On such occasions, the big-man's skill in oratory is crucial; he must be persuasive and forceful in defending his cluster's territory. His superior knowledge of genealogy and previously established land rights enables him to mediate disputes over land and other property within the cluster's territory. Negotiation with government officials concerning either the lease or sale of land is another responsibility. Some people argue that he can also forbid another Kapanaran who cannot trace descent to the cluster from gardening within the cluster's domain, though I could not discover any such restriction in the past. The retired big-man of To'ukena stated that he had never forbidden anyone from planting in his cluster's territory because "we are all from one place," highlighting the importance previously attached to village unity and solidarity. The change in some villagers' attitudes is a function of the perceived increasing economic value of land. For example, people fear that if the government purchases a part of their cluster's land, they will not receive much money if many Kapanarans from the other clusters have also established land rights in the tract being sold; those from the other clusters would also demand a portion of the proceeds of the sale.

The big-man is the spokesperson on land matters for the cluster as a whole. The senior active member of the constituent patrilineages also acts as spokesperson and guardian for his patrilineage's estate. Called the *bata ora*, or "ground father," his role is to ensure that others do not seriously encroach on the lineage's domain. He usually shows considerable tolerance in allowing anyone to prepare gardens on his lineage's strip, but if several non-cluster members intend to prepare rather large gardens together on the tract, he will protest vigorously. Such protests are usually heeded, because the gardeners fear that the *bata ora* will perform sorcery on their gardens.

At other times, disputes over property rights are not the concern of groups but of their constituent members. Ultimately, individuals are responsible for the defense of their own property rights, either by force or by persuasion. Disputes over property rights are now often settled in the village court.

Individuals, both male and female, have various types of property rights in land and other assets.[9] To describe these, it is useful to employ part of Epstein's (1969) typology of property rights with certain modifications to more accurately describe the Kapanaran tenure system.

Rights held by all members of the village community regardless of kin group affiliation are *rights of commonality*. These are denied to people from other villages. Everyone in Kapanara can hunt and gather environmental resources on fallow land throughout the territory. Some of the more important resources in which all have rights of commonality are firewood, wood for housing and fence posts, cordage, *Imperata* grass for roof thatching, a variety of edible mushrooms, wild edible tree leaves and seeds, wild betel nut (*Areca* sp.), various edible insects, birds, mammals, fish and eels, and naturally occurring substances used for both curing and sorcery. These resources are plentiful except for some mammals

---

[9] Detailing the rights of individuals and groups will not, by itself, explain observed patterns of behavior. Sometimes people garden in areas not because they have clearly specified rights but because no one objects.

and birds. In addition, a household's pigs may forage anywhere on land not used for gardens.

A *proprietary right* is the ultimate interest that an individual can have in village property. It includes the right of beneficial use and administration and sometimes the power to transfer these rights. Proprietary rights are obtained by either one's own effort, inheritance, or gift transfer and give superior title and the right to preclude others from using the property.

When a person cuts vegetation to prepare a garden on land that has not been cultivated within memory, he is clearing *araka bata,* or "new land." When the garden is fallowed and vegetation regrowth is well developed on the plot, the site and regrowth together are referred to as *kandu*. This term also signifies proprietary rights established by first cultivation, and these rights are passed by inheritance from father to son.

Rights similar to a *kandu* can be established by cultivating a piece of land that was gardened by another person in the past if either the original gardener or his descendant (if any) are unlikely to recultivate the plot in the future. During the period before administration control, for example, families sometimes emigrated to other villages to seek refuge during warfare, and some never returned to Kapanara. Kapanarans ignore such families' previous land rights.

Proprietary rights do not lapse during the long fallow period. Villagers plant both *Cordyline* sp. (*tanket* in *Tok Pisin*) and the tree *Casuarina* sp. around the borders of gardens on high-quality arable land to signify the continuity of their interests. If a person wishes to cultivate land in which another has a proprietary interest, he should first request the permission of the right-holder, and if permission is given, only temporary use rights are conferred; the grantor still retains his proprietary rights. Occasionally, however, people do not seek permission. If the person clearing the plot has a particularly close kinship or personal relation with the man holding the proprietary right and the right-holder is unlikely to recultivate the plot because of old age or change in residence, no protest will be made. At other times those with proprietary rights object to encroachment and often drape ferns such as *Pteridium aquilinum* on a stick as a symbol forbidding use of the land.

An individual can also obtain a proprietary interest in trees of economic importance such as coffee, *Pandanus* nut, and Highlands betel nut (*Areca* sp.) by planting, inheritance, or gift. Having such an interest does not, in itself, confer rights to the ground below the trees. Thus, one may receive as a gift a grove of *Pandanus* trees, but the land upon which they grow remains the property of the grantor or his kin group. However, the *de facto* situation is often different. For example, people who have been granted temporary use rights in a portion of a garden plot cleared by another occasionally request and are granted permission to plant coffee on their portion when the garden is ready to be fallowed. Although the person who initially cleared the garden may technically have proprietary rights in the ground, he is excluded from using it because coffee trees have been planted.

Because of the intermixing of land use by members of different patrilineage clusters in the past, proprietary rights in a patrilineage's tract are held by both lineage members and outsiders. Nevertheless, the patrilineage strip is associated primarily with a particular patrilineage or group of lineages and not the outsiders who have proprietary rights in its land. Previously uncultivated land remains part of the patrilineage's estate. When people established cattle projects, they did so largely on their own patrilineage's tract, even though they enclosed land in which outsiders also had proprietary rights. When the outsiders voiced objections, the cattle bosses told them that the land did not belong to them (even though they had proprietary rights) but was the property of the cattle bosses' agnatic ancestors. Such disputes underline the contingent nature of the rules of land tenure. If the cattle projects had not been built, no challenge would have been made to the proprietary rights of the outsiders.

Proprietary rights sometimes confer the right of permanent transfer.[10] A man may grant these rights in a plot that he cleared first, but cannot do so with such rights obtained by inheritance or transfer. A villager should only transfer proprietary rights in

[10] There are four types of property transfers. First, a patrilineage or cluster as a unit may grant a non-agnate a portion of its estate to reciprocate for some special service; this kind of transfer is rare today. Second, a man may sell (only to another Kapanaran, not to an outsider) for cash a plot in which he has proprietary rights, thereby transferring rights to the buyer. This occurred only twice, when men sold garden plots ready to be fallowed to other Kapanarans who wanted to plant coffee. Third, if a man is sick or old and a non-agnate continually aids him in his discomfort by bringing him firewood, water, food, or killing a pig for him, the man may reciprocate by granting proprietary rights in a plot of land or in trees of economic value. Fourth, an elder *kandere* may transfer rights to a younger one, male or female, to reciprocate for the pigs, food, beer, and help given to him by or on behalf of the younger *kandere*.

land in his own patrilineage cluster's territory and preferably in his patrilineage strip, though exceptions to this general rule can be found. Proprietary rights in trees of economic value are granted more frequently than those in land, though neither type of gift accounts for more than a very small portion of an adult's rights. Males receive these gifts more often than females, and only men are able to make the grants.

A diversity of opinion exists within the village concerning where individuals should be allowed to establish proprietary rights. In most cases, people establish these rights in the land of their own patrilineage cluster, and some villagers prefer that this practice should be the rule. Realization of the increasing economic value of land is the major factor motivating such attempts to restrict access to a cluster's resources.

Two qualifications concerning the relationship between the spatial pattern of acquiring proprietary rights and kin group membership are necessary. First, many men plant coffee trees near their homes regardless of which cluster controls the land. Second, several men reside in the hamlet of their wife's agnates, help their affinal kin in exchanges, cooperate with them in daily affairs, and garden with them. These men are felt to have "joined" their wife's cluster and thus have the right to establish proprietary rights in that cluster's land. This right passes to their sons, though these children also retain their rights in their father's original patrilineage cluster.

Rights of encumbrance give an individual a claim upon the land and assets of others by virtue of the person's relationship with the right-holders. For example, a younger *kandere*[11] may request from an older one either a tract of land or a grove of trees in which the elder *kandere* has a proprietary interest. These requests are rarely refused if relations between the two *kandere* have been good, and the transfer confers proprietary rights.

Other encumbrance rights confer temporary privileges. A woman may plant food on the land in which her father has proprietary rights and on the previously uncultivated land of her father's and mother's patrilineage clusters. A man also has the right to plant food on the previously uncultivated land of his mother's cluster, though whether this would establish proprietary rights is debat-

[11] *Kandere* is a classificatory term in *Tok Pisin* used by the Kapanarans to refer to either the mother's brothers or the sister's children. The relation between *kandere* is one of affection and mutual assistance, and the two are linked in a series of exchanges from birth to death.

able. In such circumstances, the encumbrance rights held by children of the cluster members give only the right of temporary use and not of transfer. These children's claims to land not cultivated within memory are secondary to those of male agnates in the cluster associated with the land.

Temporary use rights differ from rights of encumbrance because they depend upon the generosity of the grantor and not necessarily upon the claims of others. When a gardener has finished clearing a plot, he or she may mark specific portions of the area for others to plant food. The grant of temporary use rights is valid only for the life of the garden; when the cultivated area is fallowed, all land rights revert to the holder of proprietary rights. One person often designates three or four other villagers to garden on allotted portions, and the recipients often reciprocate with a similar type of grant within the next few years.

The duration of temporary use rights varies. In most cases, the grant enables a person to plant several times until the garden is fallowed. In other instances, the gardener who clears the plot requests that another person plant a specified crop, usually peanuts or winged beans, only once with the understanding that the grantor will be given a portion of the harvest. The grantor specifies a particular crop because his own supply is low.

With such diverse means of gaining access to land, the fundamental means of production, the traditional land tenure system was sufficiently flexible to ensure that everyone had favorable access to the resources needed to provide for the necessities of life and participation in social affairs. This flexibility was a product of conditions existing during the precontact era—chronic intervillage warfare, a relatively low population density, and the absence of commodity production. Highly restrictive rules of access to land traditionally did not exist in Kapanara, and the social relations of production concerning land control did not perpetuate economic differentiation.

CATTLE PROJECTS AND PROPERTY RELATIONS

The PNGDB was aware that no single individual within a village community controlled all the land needed to establish commercial ventures such as cattle projects (Gunton 1974: 108). Rights in land are vested ultimately in groups, and the land rights of individuals are highly intermixed. Not wishing to lend money for projects on disputed land, the Bank introduced the Clan Land

Usage Agreement. Two representatives, usually elders of the loan applicant's "clan," must indicate on the Agreement that the other members of the clan have granted the loan applicants the sole right for life to use the land enclosed by the project boundary fence for cattle raising (*ibid.*: 109). The Agreement states:

> We, the undersigned, being representatives of the . . . . Clan, hereby acknowledge that . . . . (Applicant's name) . . . . has the right under native law and custom for the whole of his lifetime to use the land known as . . . . . (or more particularly described in the plan on the reverse hereof) for the purpose of . . . . with the right to receive the proceeds of crops, trees and palms grown, livestock grazed and/or business conducted on the said land. We certify that all members of the said clan agree to the truth of this certificate and that we are the persons authorised by the clan to sign it.

To provide a description of the area referred to in the document, agricultural extension officers either make a compass traverse of the project boundary or provide a sketch plan. The Bank hopes that the Agreement will give its borrowers secure tenure to ensure that conflicting claims to land do not hinder loan repayment.

The significance that Kapanarans attach to the Agreement varies depending on the groups concerned.[12] Many consent to the enclosure even though they are not members of the cattle project. Some feel that their major gardening areas will not be affected and therefore make no protest; others hope that the project will enhance the prestige of the village. Several nonmembers do protest, but their complaints are ignored. The project followers consent to the enclosure, but they certainly do not believe that they have given up their land for their project leaders' exclusive use. They feel that they have rights in the project, and that the enterprise is on their land and not just on the bosses'. The helpers have agreed to relinquish some of their rights in the land because they expect to benefit materially from the project and to enhance the prestige of their group. In essence, the followers consent to the Clan Land Usage Agreement simply to satisfy a bureaucratic requirement, believing initially that the cattle bosses' rights will not be as encompassing as the Agreement suggests.

[12] The Clan Land Usage Agreement is not a legally enforceable document in Papua New Guinea courts. However, because government officers come to the village to obtain peoples' consent, the document has an aura of officiality and thus has influence in the village.

A brief description of a dispute between the cattle bosses and other villagers indicates that the community members never intended to grant the bosses exclusive rights. Pigs are a problem for cattle project leaders because they uproot pastures. Officials at the Kainantu Bulumakau Association suggested that project leaders demand compensation at the village court from owners of pigs that damaged cattle pastures. Some cattle bosses reported that extension officers also told them to kill such pigs. Project leaders in Kapanara threatened to make claims for compensation in their village court, and two killed a few trespassing pigs that were rooting in the pasture. Both actions made the villagers irate. At the time of the dispute, one Kapanaran scolded a boss, declaring: "It is not as if we just gave you the ground for free. You did not pay us money. If you want to shoot our pigs or take us to court, it is better that you give us money for the land." After the altercation, the project leaders ceased making claims for compensation and killing trespassing pigs. They could not gain exclusive control over use of the enclosed areas.

Nevertheless, the Agreement is important for the cattle bosses because it validates their considerable, though not exclusive, control over land enclosed by the project boundary fences. The cattle bosses stress that the land is theirs because it has been "given" to them by the other villagers, and that the boundary of the project was officially demarcated when the agricultural extension officers either surveyed or sketched the boundary of the project. They believe that the extension agents' act of surveying the boundary signifies government recognition of their claim. The bosses may also be able to secure the support of extension officers; for example, in one case an extension agent told a cattle boss to disregard the claim of another villager to land within the boss's project because the Agreement had already been signed.

Project leaders do not rely exclusively on the Agreement to validate their control over land. They assert a similar degree of control over land obtained when they expand their project even though the additional land is not covered by the Agreement. Nevertheless, the Clan Land Usage Agreement reinforces the position of the cattle bosses.

The establishment of the projects and the villagers' interpretations and manipulation of the meaning and intent of the Clan Land Usage Agreement have created two systems of land tenure. The principles specified previously apply to the area outside the projects. Although disputes concerning use of the land within the project fences exist, emerging principles can be recognized. Proprietary, encumbrance, and commonality rights have all been affected.

A villager who previously had a proprietary right or a right of encumbrance in enclosed land loses all such rights if he is not a project member. Given the previous pattern of intermixed proprietary rights, many Kapanarans have lost land rights in the fenced paddocks. Even a fellow agnate of the same minimal patrilineage unit establishing the project loses such rights if he is not a project member. The only people able to prepare gardens within a project are the members, though they also lose their proprietary rights because they cannot plant trees or transfer rights to enclosed land to others as they could outside the projects. They can only use the land temporarily. Sometimes, bosses attempt to further control their followers' use of land by restricting the number of times the garden plot can be replanted. Other bosses threaten to allow only those followers who continue working in the project to grow food in the enclosed area.

The only proprietary rights unaffected by the cattle projects are those relating to the use of trees of economic importance. Even though not a member of a particular project, an individual retains the exclusive right to use his or her previously planted trees that are now enclosed by a barbed wire fence. Only one percent or less of the village coffee is enclosed within the cattle projects. Cattle bosses forbid the further planting of coffee in the paddocks because trees remove land from grazing. In addition, because tree planting can establish proprietary rights, the project leaders view it as a challenge to their control over the land.

Rights of commonality have also been affected. Cattle bosses restrict access to naturally occurring resources within the enclosures. In the vicinity of the hamlets, most of the *Imperata* grass, which is used for thatching roofs, has been enclosed within the projects. A few cattle bosses attempt to restrict access to this resource within their project to only their followers. However, other project leaders remain indifferent on the issue because the *Imperata* used for roofing is too old and lignified to be palatable to cattle; cutting the grass stimulates new growth, which is more nutritious and palatable. Some cattle bosses allow only their followers to dig for the munah beetle grub (*Lepidiota vogeli*), which is a relished item in the diet. When the pasture is overgrazed, bosses insist that those who dig for the grub also plant improved pasture in the turned soil. Other project leaders demand only that people ask their permission before digging for the grub. The process of being asked and then granting permission, of course, creates a situation in which bosses can demonstrate generosity, thus enhancing their prestige.

Although cattle bosses do not share uniform views on limiting access to resources within their project, two general trends can be discerned. First, they are restricting the access of nonmembers. Second, they are attempting to control their followers' use of these resources, though this trend is not as marked as the first. Project leaders thus have greater control over patterns of resource use than big-men exercise.

Because the cattle projects enclose 28 percent of the grassland area, much of which is prime agricultural land near the hamlets, these changes in the social relations of production significantly influence economic differentiation in the village. In particular, the cattle bosses can use the enclosed land for other commercial purposes, such as market gardening; their followers must seek their permission first before doing so. Such permission, which may depend on a variety of factors, will not necessarily be forthcoming. What is especially important is that the project boundary fences enclose a single large area, whereas the plots in which other villagers have rights are much smaller and dispersed. If mechanization is introduced into the village agricultural system in the future, cattle bosses or their descendants will have a decided advantage in employing this technology because of the large size of their holdings.

Maintaining control over the use of land is essential for the continuation of the cattle projects and for the perpetuation of economic inequalities. According to the Clan Land Usage Agreement, the loan applicants have exclusive rights to the land only for their lifetime. However, the bosses want their sons to inherit the projects. It is difficult to forecast what will happen when all the leaders die and to what extent the competing claims of those who have traditional rights in the estate of the deceased will negate the intentions of the bosses. In the only case in which a cattle boss died, his son did assume the role of project leader and only 2 cattle of 19 in the project were killed to satisfy those who had traditional rights in the assets of the deceased. If the sons of other cattle bosses also inherit a similar position, the projects will continue to function. Such continuity in the estate would mean the inheritance of an unprecedented amount of wealth and control over resources.

CHANGING SOCIAL RELATIONS OF PRODUCTION

The introduction of smallholder cattle projects has had a dramatic effect on the social relations of production. The policies of the PNGDB and the agricultural extension officers, important components of the village's external, political-economic environment, have strongly influenced the course of change within Kapanara. These policies are concerned more with the efficient monetary output of agriculture than with the maintenance of pre-existing social relationships within the village. The focusing of external inputs almost exclusively on the project leaders and the introduction of the Clan Land Usage Agreement help to legitimate the position of the cattle bosses. The bosses have been particularly adept at manipulating their relationship with the outside world for their own gain. As a result, they have been successful in mobilizing labor without adequate reciprocation, though there are clearly limits as to how much labor followers will provide without just compensation. However, with project revenues bosses have the potential to pay others for required labor.

In the mid-1970s, economic differentiation was clearly increasing. The cattle bosses were the major beneficiaries of the projects, receiving a disproportionate share of the proceeds relative to the inputs made. Many followers felt deceived, cheated, and bitter. The pattern was not unique to Kapanara. Fleckenstein (1975: 123), for example, reported that in a project near Goroka, a similar imbalance in the distribution of benefits occurred, though the number of followers receiving significant returns was greater than in Kapanara. Complaints from followers about the failure of project leaders to share revenues adequately are widespread throughout the country. In many cattle projects, the phase of over-extraction has been reached, with many followers refusing to provide further assistance.[13]

Partly as a result of the dual system of land tenure within the village, differential access to resources has increased significantly. The traditional land tenure system's openness and flexibility, which in the past ensured that everyone had adequate access to envi-

---

[13] The ability of project leaders to maintain their enterprise after their helpers refuse to provide further assistance varies considerably in Papua New Guinea. Some hire temporary labor and successfully maintain their project. Others are unable or unwilling to do so or lose interest in the venture, and the project eventually fails. Some bosses attempt to manage their project with household labor only.

ronmental resources, has been modified markedly. Cattle bosses exercise a degree of control over a large area that is unprecedented in the village's history. They are not only restricting other villagers' access to the area enclosed by their project but are increasingly attempting to assert individual control over the enclosed land. The Clan Land Usage Agreement aids them in their endeavor. Holzknecht (1974: 71) reported a similar effort by project leaders in the Morobe Province in the lowlands:

> Once they have put a fence around a block of land, cattle project owners tend more and more to say "this is my land" and conveniently forget that permission for use of the land in this way was given by the land controller and others of his group.

In Kapanara, the control that bosses exercise over large amounts of land may be passed on to subsequent generations, thus reproducing present patterns of inequality.

CHAPTER FOUR

The Local Environment and Commodity Production

IN SOME REGIONS of the Third World, an exploitative, external political-economic environment is the main cause of increased ecological problems in rural areas. For example, many people in Kenya, Zimbabwe, and South Africa lost traditional access to much of their tribal lands due to expropriation by colonial governments and other social classes. Pressure on remaining land resources is greatly magnified in such circumstances because villagers are forced to shorten fallow periods as they cultivate smaller, more marginal areas susceptible to erosion.

However, an exploitative, external political-economic environment is not the sole cause of ecological problems. With the establishment of commercial relationships linking rural communities to outside markets, many traditional constraints on the level of local production output are removed. Historically in many tribal communities, the level of production and hence the impact on the local environment are limited by a low level of technology, a ceiling on expectations that limits demand, scheduling of subsistence activities (Flannery 1968), and institutional arrangements (Bennett 1976). In some tribal societies in Papua New Guinea, production for social and ceremonial purposes does place periodic stresses on local environmental resources, but such pressure is not continuous. Although both regional and long-distance trading links among rural communities exist in many areas, the amount produced for exchange with outsiders is traditionally small in comparison with that retained for home consumption. Households provide for most of their daily needs. Because most people produce much the same range of articles as everyone else, local demand for excess production is often limited (Forde and Douglas 1956). However, linkage to national and international commercial markets and the outside world in general removes many of these traditional limitations on output. Production can increase because the expanded marketplace often offers an unlimited outlet for local output (though producer prices may be low), and improvements in storage technology and transportation make the preservation and shipment of primary products much less of a constraint. The introduction of new, more "efficient," labor-saving technologies, such as guns or motorized transport (e.g. Nietschmann 1979), also facilitates increased output. In addition, the ceiling on expectations and on desires for obtaining material goods is removed because local production can be sold to finance the purchase of a much wider, more technologically advanced range of products than was possible in the traditional trading system. Furthermore, in Papua New Guinea the tremendous difference in the wealth of the European colonizers compared with that of the indigenous population indicates to villagers the possibility of obtaining greater wealth than before. Consequently, expectations rise. All these changes foster increased local output, leading to intensified use of resources and sometimes to environmental deterioration.

Environmental problems associated with cattle projects in Kapanara were not the direct result of oppression by external, political-economic forces. Nevertheless, government action did play a role in environmental deterioration through inaccurate assessments of the local situation. In addition, considerable environmental pressure was generated by the traditional emphasis on prestige attainment and the current *bisnis* ethos. The resulting problems of overgrazing and accelerated soil erosion in the mid-1970s not only negatively affected the potential for commodity production but also posed a threat to future subsistence production.

Similar to production systems in most peasant communities, Highlands subsistence systems are "resource-based" (see Hyden 1980: 14-16)—they depend almost exclusively on the local resource endowment, are highly subject to environmental vicissitudes, and lack the technological buffers available to more scientifically complex agricultural systems. By highlighting the nature of the local natural environment we can better evaluate the impact of cattle projects and the commercial economy on the resource base and subsistence production.

CLIMATE

As in all resource-based systems, life and livelihood in Kapanara are intimately intertwined with the pulse of the seasons. Villagers are keenly aware of such changes, basing their traditional calendrical system on various seasonal phenomena. For example, the appearance of certain birds and insects and the ripening of certain tree fruits have important significance in marking the passage of time and in indicating seasonal change. Most significant from the villagers' perspective, however, is the continual change in the sun's inclination, with the changes in seasons indicated by the setting sun's position over certain mountains. In terms of rainfall, the crucial variable affecting agriculture, people usually divide the year into three parts: *aa'u enda*, a rainy season, *ubihaero*, a relatively dry period, and *'wara tiro*, a period approximately in November. This latter period is the traditional time to plant such foods as taro (*Colocasia*), just before the big rains come.

To understand seasonality in Kapanara and the Highlands generally requires a broader depiction of regional patterns. The determinants of seasonality in the Highlands differ somewhat from the prevailing influences in the lowlands, where the pattern is directly related to the seasonal latitudinal movements of two air masses.[1] In the lowlands, from May until mid-October the southeasterly trade winds control climatic conditions, and from December until March the westerlies of the perturbation belt are the dominant influence. The two air masses are separated by the intertropical convergence zone. The remaining periods, April and mid-October to November, when the intertropical convergence zone shifts across the country, are transitional. The change to the dominance of the southeasterly trade winds is relatively rapid, whereas the transition to the controlling influence of the perturbation belt is much more gradual. The air within the perturbation belt is usually warm, moist, and unstable and has a greater capacity to produce heavy rainfall than the southeasterlies, though the latter can produce much rain where orographic effects operate (Ford 1974: 8).

The Highlands topography modifies the influence of these two air masses. Because the southeasterly trade winds reach insufficient height to pass over the mountain barrier, the Highlands and especially its eastern region are dominated externally by dry zonal easterlies from May to mid-October. During this period, local circulation, which is heavily affected by topography (Brookfield and Hart 1966: 10-11), is the major determinant of Highlands

[1] Unless otherwise noted, the description of climate is based on the work of McAlpine (1970), who in turn relied partly on the earlier research of Brookfield and Hart (1966).

weather. From December until March, the westerlies of the perturbation belt can reach sufficient height to enter and influence the Highlands weather pattern, and the perturbation-belt air mass is then co-dominant with local circulation patterns. Within the Highlands region, rainfall seasonality is more pronounced in the east, a condition that may result from the barrier ranges to the north being narrower and lower there than in other parts of the Highlands; thus, the climate in the Eastern Highlands (and hence in Kapanara) is more influenced by the markedly seasonal rainfall pattern in the adjoining Markham-Ramu lowland valley in the Morobe Province.

Long-term climatic records are available from the Highlands Agricultural Experiment Station at Aiyura (see Table 4.1), only 17 kilometers to the northwest of Kapanara. Although daily differences in weather between Kapanara and Aiyura do occur because of the highly localized nature of much of the rainfall, the general annual patterns are approximately the same. A rainy season can be discerned from December to April. A relatively dry season exists from May until September, and October and No-

TABLE 4.1

Monthly Rainfall (mm), Aiyura, 1939-1976

Month	Mean	Lower Quartile	Upper Quartile
January	229	184	266
February	269	216	319
March	258	202	322
April	218	185	266
May	124	81	158
June	89	61	115
July	98	71	130
August	121	72	175
September	126	96	176
October	156	109	205
November	190	142	239
December	239	205	280
Total	2117	1905	2310

Source: McAlpine et al. 1975:44 and data supplied by the Highlands Agricultural Experiment Station, Aiyura.

vember can be viewed as transitional between the two periods. On average, the wettest month is February and the driest is June. These long-term data, of course, mask occasional significant deviations from the monthly mean figures.

During the rainy season in Kapanara, precipitation usually starts around midday and continues on and off during the day. Night showers are also common. In contrast, the dry-season pattern is more irregular. Continual heavy downpours similar to those in the lowlands are not characteristic in the Kapanara area. An intermittent, light-to-medium intensity rainfall pattern is more common, with the stronger intensities occurring in the rainy season. During infrequent periods of very intense rainfall, minor, localized flooding occurs, and many gardens located next to streams are partially or completely destroyed.

Villagers often predict the coming of precipitation when the sky above the surrounding mountains is dark with clouds, the heat from the morning sun is intense, and a bird called *Tabibira* flies and chirps. When outdoors, they try to protect themselves from the rain with either the traditional *Pandanus* leaf mat, a plastic tarp, or occasionally a store-bought umbrella, which is considered somewhat of a luxury item.

Highlands weather is generally very pleasant, lacking the stifling and oppressive heat and humidity characteristic of the lowlands. There is little monthly variation in the daily mean minimum and maximum temperatures, with an annual mean minimum of 13.2°C and an annual mean maximum of 24.1°C (see Table 4.2). At Kapanara and Aiyura the highest mean temperatures occur during the rainy season from December to April, when temperatures during the night are kept higher than those the rest of the year because the night cloud cover traps and recycles the radiation. Although there is less cloud cover during the dry season, mean maximum temperatures are lower because of the sun's low inclination during this period. Mean minimum temperature is lowest during the dry season because the usual absence of cloud cover at night permits rapid cooling. Frost is extremely rare in Kapanara, not being the hazard that it is elsewhere at higher elevations in the Highlands (see Waddell 1972, 1975). Nevertheless, evenings and early mornings can be quite cool, and people have to rely on a smoldering fire and a few blankets to keep them warm through the night in their poorly insulated, bamboo-walled houses.

Relative humidity is highest at night, often reaching the upper

TABLE 4.2

Temperature Characteristics (°C), Aiyura, 1937-1970

Month	Mean Maximum	Mean	Mean Minimum
January	24.7	19.2	13.6
February	24.6	19.1	13.5
March	24.7	19.4	14.1
April	24.8	19.1	13.3
May	24.0	18.9	13.7
June	23.4	18.3	13.2
July	22.8	17.8	12.7
August	22.8	17.4	12.0
September	23.7	18.3	12.9
October	23.8	18.6	13.3
November	24.7	18.8	12.8
December	24.7	19.2	13.7
Annual	24.1	18.6	13.2

Source: McAlpine *et al.* 1975:124.

90s. The nearly saturated air often forms a fog in the early morning as cold air drains into the valley bottom. By eight o'clock in the morning, when the heat of the sun begins to warm up the basin, the fog disappears and relative humidity declines until approximately three o'clock, when it usually varies from the 50s to the 80s depending on the cloud cover. The relative humidity then gradually rises back to the 90s by six o'clock in the evening.

Rainfall, temperature, humidity, and wind speed (which is usually very light to moderate) all influence the rate of evapotranspiration, which in turn affects crop growth. Water stress does appear to be a problem in certain environmental zones during the dry season. For example, overgrazed pastures begin to recuperate only during the rainy season, and during the dry season certain crops such as cucumbers, corn, beans, and some leafy greens cannot be planted in gardens on the dry hillsides because they will wilt.

Water stress during the dry season has several implications. Gardening activities in certain environments are scheduled in accordance with this constraint. Also, the inability of overgrazed pastures to regrow rapidly during the dry season leaves the soil surface exposed to occasional heavy rainfall, resulting in accelerated erosion. Furthermore, limited plant regrowth accentuates grazing problems related to overstocking.

TERRAIN AND SOILS

Within the village territory, the altitudinal range is relatively limited; from the valley bottom to the highest ridge is approximately 470 meters, though the range in which hamlets and most gardens are located is only 200 meters. The topography varies but lacks the dramatic ruggedness characteristic of the terrain farther west in the Highlands and in the Southern Tairora area. Rugged, low hill ridges surround the valley, but within the basin gentle-to-medium open hills and relatively flat grassland areas predominate.

The Kapanara basin contains the northern headwaters of the Lamari River, which drains to the southern coast. Given the relatively low relief of the basin, its narrow streams are slow-moving unless swollen by very intense rains. Streams are well distributed throughout the village territory, and thus access to drinking water is never a problem.

Although villagers distinguish at least 17 types of soil and subsoil layers according to varying color, texture, friability, moisture content, fertility, and horizon, a more simplified account is sufficient here. The territory is dominated by lateritic and gleyed latosols, brown forest soils, organic swamp soils, and colluvial soils (Haantjens 1970). Most of the soils are clays, and all are acid in reaction. On the well-drained hillsides of the grasslands, the black to brown topsoil layer is approximately 15 centimeters or more deep with little humus accumulation. In some areas, iron concretions of varying size can be found at different depths, often beginning about 15 centimeters below the soil surface, giving a few hillside soils a sandy clay texture (*ibid.*: 101). On the flatter areas that are not waterlogged, the topsoil depth increases to over 30 centimeters. In the swampy areas, which are constantly waterlogged, organic peaty clay soils with a thick surface layer of decomposing vegetation can be found. Topsoil here is black, brown, or red-brown, and its depth may extend to over one meter. Gleyed horizons as well as loamy sand deposits sometimes occur below the surface. Some swamp soils are quite firm and can support the weight of a fully grown person, whereas others form an amorphous mass of soil and roots, offering little resistance to such weight. The forest soils are typical Highlands brown forest soils with a humus-rich, friable, black-brown top layer from 2 to 10 centimeters deep and a pasty, moist, brown clay layer beneath.

VEGETATION

Except in swampy areas, rain forest is the natural climax community. However, rain forest survives today only at the borders of the village territory because of past human interference through cutting and burning in the center of the basin. Forests in most cases are at a slightly higher altitude than the grasslands they enclose, a zonation that is common in the Eastern Highlands. Kapanara has two types of rain forest, mixed-oak and beech. In the west and southwest portions of the territory, *Nothofagus*, or southern beech, predominates, whereas in the north and east portions a mixed-oak forest is found. The oak forest is a two-tree-layered forest with a canopy height of approximately 30 meters (Robbins 1970: 108). *Castanopsis acuminatissima* and to a lesser extent *Lithocarpus rufovillosus* predominate in the canopy. In contrast, in the less diverse two-tiered beech forest, *Nothofagus* sp. completely dominates the upper canopy, which also forms at a height of about 30 meters.

In some areas, dense stands of the three-meter-high cane grass *Miscanthus floridulus* with a few small, scattered trees, such as *Dodonaea viscosa* and *Ficus* sp., are interposed between the forest and grassland. The community is a successional stage of garden regrowth and if left undisturbed will eventually revert to rain forest (*ibid.*: 116). However, if used repeatedly over the years, tall cane grass may remain the fallow vegetation. One section of the village territory has been used as a settlement site and garden area several times during the last 80 years. Informants state that dense forest originally covered the site, and after each period of human occupation and cultivation, the area has changed back to a forest cover more slowly. Villagers were again cultivating the area in 1976, by which time tall cane grass had become the fallow vegetation, and it is possible that such continued interference will eventually deflect the succession towards short grassland vegetation (*ibid.*).

In Kapanara, as in many Eastern Highlands villages, the dominant vegetation is induced short grassland. These grasslands are anthropogenic in origin, resulting from periodic recultivation, the cutting of trees for fuel and housing, the rooting of domesticated pigs, and burning (Robbins 1963; Brookfield with Hart 1971). The distribution of short grasses is related to slope and attendant soil moisture status. On the drier hillsides, the tussock-forming *Themeda australis* dominates the grassland areas, forming a canopy about 60 to 90 centimeters high. Mixed with the *Themeda* are the emergents *Ophiuros tongcalingii* and *Arundinella setosa* and other grasses such as *Imperata cylindrica, Eulalia trispicata, Capillipedium parviflorum, Ischaemum barbatum,* and the sedge *Fimbristylis dichotoma.* Ferns such as *Pteridium aquilinum* and *Cyclosorus unitus* are scattered through the grasslands. The under-story is made up of the small woody melastome *Osbeckia chinensis* and small forbs such as *Polygala japonica.* On less inclined slopes with deeper soils and increased soil moisture, *Imperata* and *Ischaemum* become more frequent. In the valley bottom where the soil is very moist, many of the above species are replaced by the sprawling grass *Ischaemum polystachyum,* which forms a dense cover about one meter high. Associated with this grass are *Leersia hexandra, Sacciolepis indica,* and scattered meter-high *Phragmites karka.*

Located in the valley floor, the swamp grassland community is permanently waterlogged, with water often above the soil surface. Dense stands of *Phragmites* up to six meters tall usually dominate the vegetative cover, though *Ischaemum polystachyum* and *Leersia* occasionally achieve prominence. Other common species are the erect grass *Isachne brassii,* the sedge *Cyperus melanospermus,* and the small forbs *Oenanthe javanica, Polygonum strigosum, Polygonum barbatum,* and *Commelina diffusa.*

These vegetative communities evolved without the influence of a large grazing animal, but have now been modified considerably under moderate-to-high grazing pressure. Within the project boundary fences, villagers have enclosed mainly swamp and induced short grassland vegetation.

THE IMPACT OF GRAZING ON THE VEGETATION

The impact of grazing on ecological systems has been the subject of worldwide interest, and a voluminous literature on the subject exists. In overstocked pastures, cattle graze out the palatable, accessible perennial species; unpalatable[2] perennial and annual plants and recumbent species, which may not be accessible to cattle, often thrive. Because of their reliance on seeding for survival, annuals are able to colonize bare patches as grazing reduces the vegetative cover. Invasions of brush and small trees result from the reduced frequency of fires as less dried vegetation is available to burn. Net primary productivity of the grassland is reduced in overgrazed pastures (Vickery 1972: 312), trampling damages many plant species (Edmond 1966: 453), and thinning the vegetative cover creates the potential for greater erosion.

Frequent and intense grazing reduces the size and growth of the plant roots, and the greater the intensity of grazing the more weakened the root system becomes (Alcock 1964: 27). Pre-existing roots may begin to decompose. The reduced root system restricts the plant's access to water to the top layer of the soil, and it is less able to absorb soil water held at a high moisture tension; both changes become especially harmful in a place where dry spells normally create water stress in plants, as in the Eastern Highlands.

In perennial plants, some carbohydrates manufactured in photosynthesis are used for growth and maintenance, and the rest are stored as reserves in roots, rhyzomes, and leaf bases. When the plant is defoliated, it can no longer adequately supply the roots with reserves, and the root reserves are eventually depleted through root respiration. As the rate of root respiration declines, so does nutrient uptake, and thus the plant suffers (Davidson 1968: 131).

The removal of part of the leaf cover through grazing alters the microclimate, which influences interspecific plant competition. With the removal of the top layer of vegetation, more radiation reaches the soil giving a new competitive edge to those low-growing species normally shaded out when a full vegetative cover exists. A more arid microclimate is also produced as evaporation increases (Liddle and Moore 1974: 1057). Water infiltration decreases, and more surface runoff occurs after rain because the much-reduced mulch cover no longer retards the overland flow of water. Another microclimatic change that is related to the decrease in the protective cover is the increase in temperature extremes at the soil surface.

[2] The term "unpalatable" must be used in a relative sense. When available pasture is reduced, cattle will eat vegetation that they previously avoided.

Some researchers (e.g. Chappell *et al.* 1971: 880) have also stressed the importance of trampling in bringing about vegetation change. In a mature bovine, the weight under each hoof is approximately 3.2 kilograms per square centimeter (Duffey *et al.* 1974: 182), and this weight both compacts the soil and damages certain types of plants. Thus, treading favors those species more adapted to compacted soils and with morphological features better able to withstand the pressure of treading.

Despite all these potential problems, cattle grazing *per se* is not necessarily deleterious to the environment. Indeed, with carefully controlled grazing the net primary productivity of the pastures can be greater than the rates for ungrazed areas (Vickery 1972: 311). Grazing can reduce the leaf area index of plants to an optimum level by preventing self-shading, and it also stimulates the production of new tillers (Kumar and Joshi 1972: 673). In addition, old leaves are less efficient in photosynthesis than young ones, and carefully controlled grazing can prevent the survival of much old leaf tissue.

Such carefully controlled management necessary to ensure the maximum net productivity of the pasture and to minimize the potential for environmental damage is lacking in village smallholder cattle projects for a variety of reasons. First, knowledge concerning proper cattle husbandry and the effects of poor management is limited because villagers have no tradition of cattle raising. Most knowledge influencing herd management is based on information obtained from agricultural extension agents, from the rudimentary training that one or two individuals receive at government-run cattle husbandry schools, and from other villagers involved in cattle raising. Second, very often those who attend government-run schools for cattle husbandry are too young to have the status necessary to influence either the cattle bosses or their followers to provide quality management. Third, knowledge alone will not ensure high standards of management. Although planting improved pastures, reducing erosion potential, and other management inputs are environmentally sound, these activities will not, by themselves, augment an individual's prestige. Villagers, however, are more concerned with obtaining prestige. Cattle projects do not exist as entities separate from the cultural milieu but are aspects of social strategies designed to gain money, prestige, and power. The more stock within a cattle project, the more successful the endeavor is considered and thus the greater the prestige accruing to the people controlling the project.

The Kapanara project that had the most stock in the mid-1970s was seriously overgrazed, but all felt that it was nevertheless the most prestigious. Thus, environmental pressure is generated by the traditional cultural emphasis on prestige attainment in the context of commodity production. Fourth, a high standard of project management requires more time than villagers are willing to allocate; other activities, such as subsistence gardening, coffee production, visiting, or leisure, compete for time. Finally, even when project leaders do want to sell some cattle, which would reduce grazing pressure, difficulty in marketing during certain times can discourage them from doing so.

The extent of environmental problems in Kapanara caused by grazing is related both to the stocking rate and to the system of grazing employed. Excessive stocking rates exceeding the carrying capacity of the range result in overgrazing and erosion, and the system of grazing management can exacerbate existing pressures.

As in most smallholder projects in Papua New Guinea, agricultural extension officers determined the initial stocking rates in Kapanara. When they processed loan applications, they suggested that the PNGDB lend money for purchasing an initial herd that was below the carrying capacity of the enclosed pasture because the herd would increase as calves were born. Stocking rates were based on certain assumptions about changes over time in the age and sex ratio of the herd, in calving and mortality rates, in the number of cattle sold, and in the proportion of natural to improved or semi-improved pasture; the more improved pasture planted, the higher the stocking rate because such pasture is much more nutritious than native pasture. In the Kainantu District, extension officers usually set the stocking rate on natural pasture at one animal to 2.5 hectares and on improved pasture, at higher densities.

Because the last four projects established in Kapanara had not been fully stocked in the mid-1970s, it was only possible to examine the oldest three projects in terms of how accurate estimates made by agricultural extension officers were in relation to stocking rates and the area of improved pasture planted. The number of animals in these three projects in 1977 was from 24 to 46 percent less than the number originally predicted by the agricultural extension agents because the number of sales and the death rate had been higher than originally expected. Also, members of two projects had expanded their enclosed area by 1975, thus further reducing the stocking rate. Counterbalancing these factors, villagers planted much less improved pasture than the agents recommended, and consequently the carrying capacity of the projects had been overestimated. Improved pasture did not cover more than 5 percent of the total area in any project, whereas extension agents estimated that approximately 30 percent of the pasture would be improved. However, because cattle bosses of two projects expanded the area of their enterprise and kept less stock than extension officers envisioned, overgrazing did not occur. If the two projects had had stocking rates of one animal per 1.75 hectares as originally recommended but without the amount of improved pasture expected, serious overgrazing would most likely have resulted. For the third project, a high stocking rate of one animal per 1.14 hectares had been suggested; without much improved pasture such a rate would have been disastrous. However, by late 1977, the project had only 44 cattle—less than anticipated—giving a stocking rate of one animal per 1.94 hectares. Nevertheless, clear signs of environmental degradation—overgrazing and accelerated erosion—were evident. Such faulty estimates by some extension officers were thus partly to blame for overstocking, though the situation was much more complex. Certainly, poor management was also a contributing factor.

The form of grazing management employed influences the degree of environmental disturbance. Two major systems of management are continuous grazing and rotational grazing. In the former, cattle graze wherever they wish in an enclosed area. The advantage of the system, which works well with low stocking densities, is its low cost and low level of required management. In the latter system, small fenced subdivisions are built within the paddock, and cattle are allowed to graze in one subdivision for a short time, thus creating artificially high stocking rates. Before grazing out any pasture species, the cattle are supposed to be moved to the next subdivision. After the grass within a small subdivision has regrown, cattle are then brought back to graze. The advantage of rotational grazing is that it forces cattle to graze the pasture uniformly instead of allowing them to graze out areas with more palatable species while ignoring the rest of the pasture, which would become rank. Rotational grazing has other advantages in controlling the number of parasites and in favoring improved pasture species, which recover from grazing faster than most native grasses (E. Henty, personal communication: 1979). However, rotational grazing is expensive because of the cost of wire for subdivisions, and it requires a high level of management, which in most cases is lacking. If cattle are allowed to stay too long in one small subdivision, overgrazing will occur, the animals will lose condition, and the rest of the pasture will become rank.

In the early and mid-1970s, agricultural extension officers in the Kainantu District persuaded villagers to install subdivisions for the purpose of rotational grazing. Nevertheless, in many villages rotational grazing was not practiced because subdivision gates were usually left open, and the wire on subdivision fences was often loose. In Kapanara, when people did attempt to use rotational grazing, they were generally unsuccessful. For example, in the heavily overgrazed project, a small subdivision was fenced off to allow grass to regrow. The area was closed off for too long, and consequently the pasture became rank and the cattle would not graze it. The cattlemen thus had to burn it so that new, more nutritious and palatable pasture would grow. At the same time, the rest of the pasture was subjected to an abnormally high stocking rate, which facilitated overgrazing.

In effect, continuous grazing is generally practiced in Kapanara and most other Highlands villages. The major disadvantage of this system under moderately high stocking rates is that once an area has been grazed down cattle often return because all the plant material is new, tender, and more palatable. Eventually, the area is grazed out and invaded by weeds, while other areas are allowed to become rank and have to be burned to provide palatable pasture.

Cattle were not the only agents of environmental modification. Agricultural extension officers attempted to persuade villagers to plant semi-improved pastures of legumes and introduced nutritious grasses, which help cattle gain weight more rapidly and make higher stocking rates possible. Legumes are particularly important because some species can fix atmospheric nitrogen and thus add nitrogen to the soil, thereby improving the nutrient supply to grasses in the pasture. Nitrogen content is especially critical in tropical pastures. Weight gains in cattle are determined by how fast food consumed passes through the rumen, and the speed of this process is affected by the digestibility of the grasses and their nutrient content. If the crude-protein level of the diet falls below 8.5 percent, rumen activity and hence the amount of food eaten may decrease because of a lack of nitrogen to supply the micro-organisms that help digest food (Milford and Minson 1966: 108). Cattle on tropical pastures show a marked decline in their food intake when the crude-protein proportion of their diet falls below 7 percent (*ibid.*). Not only are tropical grasses low in digestibility because of a high crude-fiber content, but many native grass species have a crude-protein level below 7 percent when mature. These deficiencies apply to both *Themeda australis* and *Imperata cylindrica*, major components of native grasslands in the Eastern Highlands (Schottler 1977: 4). Unlike native grasses, introduced legumes maintain a high crude-protein percentage and a high level of digestibility when mature. Introduced, improved grass pasture species also have a higher protein content than native grasses.

Because of these advantages, agricultural officers distributed to villagers improved species of grasses and legumes, the latter having approximate crude-protein levels from 15 to 20 percent (*ibid.*). To offset the expense of purchasing pasture seeds, the PNGDB usually provided funding for pasture improvement in the original loan grant.[3] However, official government enthusiasm for improved pasture was not matched by a corresponding village effort to plant these species. Not only did Kapanarans poorly understand the relationship between improved pasture and the rate of weight gains in cattle, but they also felt that planting a significant portion of the pasture with improved species was too time-consuming and expensive.

Within Kapanara, the most commonly planted introduced grass and legume were *Setaria anceps* (setaria) and *Stylosanthes guyanensis* (stylo) respectively.[4] The former was usually planted vegetatively, whereas the latter was sown onto burnt grassland. Within a few projects, villagers established small (less than 100 square meters) nurseries of *Setaria*, but the nurseries were often short-lived and not replanted. Planting practices were quite casual, never intensive, and sometimes were nothing more than tossing a few seed heads of *Setaria* near footpaths where pigs had turned the soil.

Pigs have also been a significant factor in bringing about environmental change within grazing areas. The pig's major foraging

[3] The range in the amount of money allocated in the loans for pasture improvement in the seven Kapanara cattle projects was from zero to K200. The cost of improving 50 hectares by oversowing all the area with the legume stylo in 1977 was about K500 (Gutteridge 1977: 1).

[4] Villagers planted other improved pasture grasses, such as *Brachiaria mutica* (para grass) and *Pennisetum clandestinum* (kikuyu grass), and other legumes, such as *Desmodium uncinatum* (silver leaf desmodium) and *Calopogonium mucanoides* (calapo), but the areal extent of these was quite small. Some species of improved pasture such as *Pennisetum purpureum* and *Pennisetum clandestinum* can become serious garden weeds (Graham 1973: 62).

area is the swamplands, but when rooting on well-drained hillsides they prefer areas already heavily grazed as opposed to hillsides having a full grass cover. Their rooting loosens the soil and leaves patches of bare ground, thus facilitating erosion and creating a favorable environment for weed invasions. Without drawing any definite conclusions, it is possible to speculate on the reasons for pigs preferring overgrazed areas. One possibility is that the pigs, desiring food with a high protein content, seek the high protein nodules of the introduced legumes and the roots of legumes and introduced grasses; however, this argument does not provide a complete explanation because the areal extent of pigs' intensive rooting is greater than the area under improved pasture. More likely, grazing may bring about environmental changes in the vegetation and soil that favor certain species of insects within the soil; Roberts (1978), for example, found that different insect populations thrive at different stocking densities under sheep grazing. In Kapanara, people asserted that more larvae of *Lepidiota vogeli* (munah beetle) occurred within the heavily overgrazed project than elsewhere within the village territory, a difference possibly related to intensive grazing. Pigs eat the larvae and may thus prefer the overgrazed area because of the greater incidence of *Lepidiota* there. Another possibility is that the roots of the invading weeds are more palatable for pigs than the roots of the natural grassland cover (J. Giles, personal communication 1978). Whatever the cause, the impact of cattle is exacerbated greatly by pigs preferring to root in areas already overgrazed.

To examine grazing-induced vegetation changes, I compare results from a series of vegetation transects on both the dry hillsides and the swampy areas in the heavily overgrazed project with data from other transects in either very lightly grazed or ungrazed areas elsewhere (see Appendix A for methodology). Differences in the percentage of leaf cover and the frequency of each species within a meter-square sampling frame moved along the transects reflect the impact of cattle. Changes in species frequency imply changes in cover, but the converse is not true. In addition, the combined percentage leaf cover of all species underneath the transect lines is an indicator of the potential for accelerated erosion.

In the overgrazed pasture on the drier hillsides, the original 50 to 60 percent cover of *Themeda* was gradually thinned to a range from 50 percent to less than 25 percent. Other palatable native grasses such as *Ischaemum barbatum*, *Capillipedium parviflorum*, and to a lesser extent *Ophiuros tongcalingii* also decreased in frequency, and the areal coverage of *Imperata cylindrica* was slightly reduced. Cattle grazing also lowered the frequency of such native legumes as *Pueraria triloba* and to a lesser extent the palatable annual composite *Crassocephalum crepidioides*. Grasses such as *Eulalia trispicata*, *Arundinella setosa*, and *Sacciolepis indica* appeared unharmed by grazing. A host of weeds took advantage of the patches of bare ground and increased light penetration through the vegetative cover. Such creeping forbs as *Hydrocotyle javanica* and *Oxalis corniculata* and the short, creeping grass *Digitaria radicosa* became more common. The erect perennial herb *Erigeron sumatrensis*, which grows as high as two meters, became more frequent particularly in those areas disturbed by pigs. The annual composite *Ageratum conyzoides*, one of the most successful invading weeds in the overgrazed area, sometimes achieved a leaf cover of over 25 percent. Other grasses such as *Eragrostis tenuifolia*, which is common in compacted soils, *Digitaria violascens*, *Paspalum orbiculare*, the annual subshrub *Urena lobata*, and the short, annual, white-flowered forb *Polygala paniculata* all increased in frequency.

In the moist-to-swampy soils in the valley bottom, the rate of vegetation change was much faster in part because cattle preferred to graze in the lush pasture there before moving to the hillsides. On the dry hillsides, useful pasture, although reduced in percentage cover, still survived, whereas in some swampy areas, weeds dominated completely.

In moist-to-swampy soils, the major change was a reduction in the cover and frequency of two previously dominant grasses, *Ischaemum polystachyum* and *Phragmites karka*, the former being especially preferred by cattle. The leaf cover of each, which was previously up to over 50 percent, declined to less than 25 percent. Project members had either cut or burnt the pre-existing *Phragmites* to provide new, more tender growth for the cattle. This grass was never able to regain its former position of predominance and remained cropped short because of continuous grazing pressure. The grass *Isachne brassii* and the forb *Pouzolzia hirta* also decreased in frequency. Some grasses of the original pasture, especially *Leersia hexandra*, were not affected adversely by grazing pressure. The sedge family (*Cyperaceae*) was the main group increasing in occurrence in the overgrazed swampy areas. In some places, dense stands of either the 1.3-meter-high *Fimbristylis salbundia* or the slightly shorter *Eleocharis dulcis* predominated. Other sedges such as *Cyperus pilosus*, *Eleocharis tetraquetra*,

Cyperus globosus, Cyperus melanospermus, and *Cyperus distans* also became more frequent, but to a lesser extent. The sub-shrub *Ludwigia octovalvis,* which can cause gastroenteritis in cattle (Henty and Pritchard 1975: 130), the annual grass *Echinochloa colonum,* the semi-erect perennial herb *Polygonum barbatum,* and the short herbs *Limnophila aromatica* and *Lindernia antipoda* also increased in occurrence. The composite *Ageratum conyzoides,* which was found on the dry hillsides, also became more frequent in swampy as well as in moist, compact soils. Other plants such as the short, edible, but slow-growing grass *Paspalum conjugatum,* the very strongly rooted perennial shrub *Sida rhombifolia,* and the one- to two-meter-high perennial forb *Verbena bonariensis* became more common in moist, compact soils.

More important from the perspective of increased erosion potential is the decrease in the total vegetative cover on the drier hillsides. In the ungrazed and very lightly grazed areas, the vegetation cover under the line transects was 88 percent, whereas in the overgrazed area, which exhibited much more variability in terms of cover, the approximate leaf cover was from 50 to 60 percent. Several factors account for the lower figure for the overgrazed hillsides. Although a high percentage of leaf cover is evident in relatively ungrazed areas, the basal cover there is much less. When overgrazing occurs, the height of the palatable species is reduced, exposing the soil beneath where no plants have rooted, though an increase in weeds partially covers the spaces between the plant bases. Also, cattle trails and pig rooting leave bare patches in the transects. In addition, overgrazing weakens pasture growth.

Overgrazing was thus a serious problem, a condition I observed in many smallholder projects throughout the Highlands in 1977. The cattle within the heavily overgrazed project were thin and scrawny compared with those in the newer projects, where grazing pressure was much less intense. When the amount of quality herbage was severely reduced, the cattle bosses could either let the cattle starve, sell off a considerable portion of the herd, or expand the paddock area to enclose a greater food supply for their animals. The first and second alternatives were considered undesirable, because controlling a large herd brings prestige. Therefore, the people involved in the overstocked enterprise expanded the size of their paddock in late 1977, just as members of two other projects did previously. Before doing so, some members occasionally herded their cattle on open grassland because of the pasture shortage. The expansion of the project areas, in turn, has serious implications for the agricultural system, which will be examined in the next chapter.

THE IMPACT OF GRAZING ON THE SOIL

Overgrazing can produce a more accelerated rate of erosion than traditional land-use patterns. Although garden sites are exposed to erosion when new, the gradual growth in the crop cover and the short period of cultivation reduce the long-term environmental impact. In contrast, grazing can continually leave the soil surface exposed to the erosive force of the rain.

Before erosion can occur, soil particles must be detached from the soil mass. Several forms of disturbance can accomplish this. Pigs rooting in overgrazed pasture is one means. Treading by cattle is especially effective in loosening the surface layer on hillsides in the dry season. The major force is the direct impact of raindrops on the soil surface, and thus, the greater the reduction in vegetative cover, the greater the erosion potential. Raindrops hitting the soil-water surface create turbulence and break down and disperse the soil aggregates. The splashing that results enhances sediment transport. A thin film of sediment forms on the soil surface which, in turn, reduces the infiltration rate[5]—a sealing effect that is most pronounced in clay soils. Faster overland flow results, enabling more sediment to be carried downslope.

Soil condition also influences the infiltration rate and erosion potential. Clay soils, such as those in Kapanara, have the greatest potential for compaction, which reduces the rate of infiltration and the soil's water-holding capacity. Soils in overgrazed areas are drier, and drier soils permit more rapid rates of infiltration, but such an effect is countered by increased compaction. Certain conditions in the tropics facilitate the detrimental impact of compaction. In tropical pastures as in Kapanara, plants do not provide a full basal cover, and bare spots between plant bases can be more easily compacted (Russell 1966: 39). Also, in temperate regions winter frost heaves the soil surface, thus partly alleviating the impact of compaction, but no such action occurs in the tropics except in a few isolated, high altitude regions (Thomas 1960: 92).

Although compacted soils are more closely bound together,

---

[5] The thin film also forms on sweet potato mounds. Villagers scratch the surface of the mounds after heavy rains to remove the crust on the clay soil and to increase the infiltration capacity of the mounds.

they have a greater potential for erosion because of the increased amount of runoff (Meeuwig and Packer 1976: 112).

> Although overland flow is not very effective as a soil-detaching agent unless concentrated in rills or gullies, the presence of overland flow increases the effectiveness of raindrops as detaching agents. If the soil surface is saturated and covered with a film of water, the raindrops striking it tend to rebound and to dislodge and transport soil particles. If the soil is only moist, the raindrops tend to penetrate the soil, dissipating at least part of their kinetic energy as frictional losses. (*ibid.*)

A comparison of soil bulk density data obtained in the field reveals the extent of cattle-induced soil compaction. Test sites were chosen inside the heavily overgrazed cattle project and inside an adjacent, relatively ungrazed project with a low stocking density of one animal per 8.5 hectares. To make controlled comparisons between the two areas, test sites in each area were matched as closely as possible in terms of the degree and length of slope, vegetation type, and soil texture (see Appendix A).

As expected, soil bulk density of the top 5 centimeters was higher in the overgrazed area (see Table 4.3). Cores taken near the stockyard, where cattle in the overgrazed project often congregated, had the highest mean bulk density (1.00 gram per cubic centimeter).[6] In the overgrazed project, soil bulk densities were 5 to 40 percent higher in the well-drained hillsides and 29 to 44 percent higher in very moist soils near the swamp edges.[7]

The data on both bulk density and vegetative cover suggest increased erosion potential. Even a cursory inspection of the overgrazed paddock revealed that accelerated erosion affected the area in the mid-1970s. Loose, unconsolidated soil was trapped at the base of plants on the up-slope side; in one small area minor rills were cutting the soil surface; and in the depressions caused by hoofprints small soil particles were trapped after being trans-

---

[6] This figure is low for clay soils and results from either a high swelling capacity when wetted, a high level of organic matter in the soil, or volcanic ash being present (McIntyre, personal communication 1978).

[7] Wetter soils usually have a greater potential for compaction, and two other sampled sites in the moist soils within the overgrazed project did have much higher bulk densities, but the data had to be rejected for various reasons. Perhaps with a greater number of sample sites in the very moist soils, a higher degree of compaction would have been found in the overgrazed area. The difference in soil bulk density in the well-drained hillsides and at the edge of the swamps in the two areas was statistically significant to greater than the .001 level.

TABLE 4.3

Bulk Density of Top 5 cm of Soil (g/cm³)

Group Number	Site Description	Relatively Ungrazed Area		Overgrazed Area	
		Samples	Mean of Samples	Samples	Mean of Samples
1	Stockyard	.86	.87	.90	1.00
		.86		1.00	
		.88		.98	
				1.12	
2 (Well-Drained Hillsides)	Top of hillcrest	.64	.68	.90	.92
		.66		.98	
		.75		.88	
		.68			
	Top of hillcrest	.78	.77	.84	.81
		.84		.82	
		.70		.76	
		.74			
	43 m from hilltop	.67	.70	.89	.88
		.74		.87	
		.72		.87	
		.65			
	45 m from hilltop	.72	.74	.82	.90
		.73		.92	
		.86		.96	
		.65		.91	
	50 m from hilltop	.79	.78	1.00	.92
		.79		.86	
		.77		.89	
		.78		.93	
	58 m from hilltop	Samples lost		1.01	.98
				.97	
				.96	
	75 m from hilltop	.75	.74	.98	.97
		.72		.99	
		.75		.98	
				.94	

TABLE 4.3 (*cont.*)

Bulk Density of Top 5 cm of Soil (g/cm³)

Group Number	Site Description	Relatively Ungrazed Area		Overgrazed Area	
		Samples	Mean of Samples	Samples	Mean of Samples
	Edge of swamp	.59	.59	.81	.78
		.54		.81	
		.60		.69	
3		.61		.81	
		.59			
	Edge of swamp	.53	.54	.75	.76
		.52		.71	
		.52		.75	
		.58		.84	

Group 1: $t = 2.90 > 2.35 = t_3$ DF; $p < .05$, 1 tail

Variance Ratio $= 62.0 > 19.2 = F_{3, 2}$ DF; $p < .05$

Group 2: $t = 9.83 > 3.29 = t_{45}$ DF; $p < .001$, 1 tail

Variance Ratio $= 1.24 < 2.03 = F_{23, 22}$ DF; $p < .05$

Group 3: $t = 9.35 > 3.73 = t_{15}$ DF; $p < .001$, 1 tail

Variance Ratio $= 2.20 < 3.50 = F_{7, 8}$ DF; $p < .05$

ported either by runoff or splashing. Measurements of soil accumulated against calibrated planks set in various parts of the slopes in the overgrazed area and the adjacent, relatively ungrazed project confirmed the visual impression of accelerated erosion. The level of soil accumulation was over 15 times as much in the overgrazed zone, whereas very little soil collected against boards in the relatively ungrazed area (see Appendix A). This degree of difference in erosion is overly conservative because certain factors in the field situation, such as the presence of cattle trails, hindered the collection of data on the full extent of soil loss in the overgrazed area.

For people who depend on the land for their livelihood, accelerated erosion can be a serious problem because it can reduce agricultural potential, though detrimental results from erosion are not inevitable. A problem in determining the significance of erosion is that the loss of agricultural productivity is not directly proportional to the amount of sediment collected because many nutrients may be dissolved in the surface runoff and not trapped by the boards (Leopold 1956: 640). Also, the impact of erosion depends on the depth of topsoil and the rate at which topsoil can be replaced through weathering. The soils on the Kapanara hillsides are generally thin, and the rate of weathering is slowed by the lack of rain in the dry season and by generally cool temperatures for a tropical zone. Given these considerations and the large difference in erosion in the two test areas, it is possible to conclude that accelerated erosion within the overgrazed area was reducing the agricultural potential of the eroded land in the mid-1970s.

In assessing the impact of erosion on the agricultural system, the uses of the drier hillsides must be specified. Some land under *Themeda* grassland is not valued for gardens unless settlement sites are nearby, in which case the low yields from these areas are offset by the short distance to the settlements. Much land near the hamlets is enclosed within the cattle projects, and in such areas *Themeda* grassland would be cultivated. In addition, where patches of *Imperata* and bracken fern occur on the hillsides, the land is valued for yam gardens. Thus, erosion was definitely affecting land suitable for cultivation.

Soil compaction also poses a problem for the agricultural system. By compacting the soils, cattle grazing negates any positive effects of the fallow period on bulk density. Fallowing, besides allowing a build-up in available nutrients, is necessary to restore the soil structure; old gardens, for example, have a higher soil bulk density compared with fallowed areas (Clarke 1971: 73; Scott 1974: 59). Villagers, in appraising the present agricultural utility of the compacted soils along the swamp edges, asserted pessimistically that if they cultivated there now, yields would be abnormally low; such locations, however, are usually prime gardening areas.

What was particularly disturbing in the mid-1970s was that *any* serious environmental problems had occurred at all. Indeed, the enterprise with the heavily overgrazed pasture had had cattle for only three years! An ecological view requires a consideration of the long-term, potential environmental impact, and similar overgrazing over long periods would have serious implications for land-use practices and the future of the Kapanarans' livelihood.

CHAPTER FIVE

Subsistence and Commodity Production

> A subsistence system can be thought of as the complex of functionally related resources and activities through which a group secures food for its own needs and by its own efforts, usually by the direct exploitation of its environment. (Nietschmann 1973: 2)

TRADITIONALLY, the subsistence system of Kapanara was not only a food procurement system but also a culturally and socially meaningful complex of activities through which interpersonal relationships were established and reinforced with both the living and deceased. As in the past, agriculture and pig husbandry are the main components today, supplemented by hunting and gathering and arboriculture. The maintenance of a viable subsistence system is absolutely essential for the welfare of the people, because Kapanarans, like other peasants, depend largely on local food production for their sustenance.

The conventional wisdom asserts that if surplus land and labor exist in rural communities, the introduction of commodity production will not conflict with subsistence endeavors. In Kapanara, a surplus of total land and labor was available, but nevertheless the introduction of commodity production seriously undermined local food production in the mid-1970s. To understand the impact of cattle raising on subsistence production, we must broaden our analysis to include not only total land and labor available, but also (1) the nature of the linkages within the subsistence system, particularly those between food gardens and pigs; (2) the relative location of subsistence and cash-earning enterprises; and (3) the community's sociocultural characteristics, especially the emphasis on prestige attainment and the *bisnis* ethos.

THE PRESENT AGRICULTURAL SYSTEM

In terms of the percentage of calories contributed to the diet, agriculture is clearly the most important subsistence activity. The area under subsistence agriculture is approximately 0.08 hectare per capita.[1] Only human labor and fire are used in production; no draught animals or machines are employed. The agricultural tool kit consists of the steel axe, spade, bushknife (a machete), and wooden digging stick. Villagers recognize the importance of hard work, favorable weather, and ritual in promoting productive gardens. Except for a few tasks, the division of labor by sex is not strict. Men fell large trees, fence gardens, dig drainage ditches, and for ritual purposes care for fully grown banana plants and sugar cane. Women largely, but not exclusively, clear grass, plant, harvest, and weed. Both men and women are involved in tilling the soil. Occasionally, husband and wife cooperate and work together on such tasks as clearing grass or planting.

In order to understand the dynamics of subsistence production as it becomes intertwined with commodity production, we must first characterize the diversity in the agricultural system. As Boyd (1981) notes for the Awa people to the southwest of Kapanara, any assumption of homogeneity in subsistence agriculture is invalid. Diversity exists at two levels. First, villagers prepare several types of gardens in different environmental zones. Second, they vary not only in their preference for different types of gardens but also in their style of gardening within the same environmental zone. Such diversity among individual cultivators is characteristic of many subsistence systems and is an important basis of experimentation and change in agricultural systems (Johnson 1972).

The emic classification of garden types cannot be reduced to any simple formula. Sometimes villagers distinguish subsistence gardens by the crop complexes within them. They also make contrasts by adding the suffix kai'a ("work") to a term that signifies either some characteristic of the site or the environmental zone in which the garden is found. People do not always agree on the particular term that should be used, though differences in folk taxonomies are to be expected (see Hays 1974).

When classifying a garden by the crop complex within it, people

---

[1] This figure is based on a compass and tape survey of the gardens of 13 households. The term "subsistence" normally refers to production for use as opposed to production destined for exchange with those outside the household. I use "subsistence" to refer to three types of local food production: (1) that produced and consumed by the household; (2) that given to other households through reciprocal and ceremonial exchanges; and (3) food crops that are largely consumed at home—such as sweet potato, corn, and taro kongkong—but occasionally are also sold in the town market. Commodity production in the village involves mainly coffee production and cattle raising. See also Brookfield (1972) for distinctions among different types of production in Papua New Guinea.

add the suffix *naaho* ("garden") to the name of the dominant crop or crops (see Table 5.1). Each garden type displays a particular spatial arrangement of crops that sometimes varies according to the environmental zone. In grassland sweet potato gardens, for example, sweet potato, maize, pumpkin, and beans are normally found in the center, while banana, sugar cane, tobacco, and such greens as *Setaria palmifolia* and *Rungia klossii* are located along the perimeter, but in forest sweet potato gardens the layout is less patterned. Another type of garden, the household or dooryard garden, is called a variety of names such as *batuka naaho* or "place garden" and is often composed of a variety of crops displaying no orderly arrangement. Villagers also plant some sites largely with "cash crops" such as peanuts, cabbage, pineapple, or tobacco, though such plots cannot easily be distinguished from subsistence gardens because a large percentage of these crops is often not sold but consumed by the households and other subsistence crops are interplanted with them.

As is true in most Highlands communities, the sweet potato garden is clearly the most important crop complex in Kapanara (see Table 5.2).[2] Within these gardens other crops can also be found, yielding a succession of produce to harvest (see Table 5.3).

TABLE 5.1

Major Garden Types

English Name (Botanical Name)	Tairora Name
Sweet potato (*Ipomoea batatas*) garden	*aama naaho*
Taro kongkong (*Xanthosoma* sp.) and taro (*Colocasia* sp.) garden	*tapae-kara naaho*
Taro, taro kongkong, and yam (*Dioscorea* spp.) garden	*kara-tapae-oba naaho*
Taro and yam garden	*kara-oba naaho*
Winged bean (*Psophocarpus tetragonolobus*) and yam garden	*ae'i-oba naaho*
Sugar cane (*Saccharum officinarum*) garden	*ka'a naaho*
Various leafy greens gardens such as a *Rungia klossii* and *Setaria palmifolia* garden	*no'u-bwaaki naaho*

[2] To determine the areal extent of the different garden types, I surveyed with

TABLE 5.2

Percentage of Total Cultivated Area under Different
Crop Complexes, Sample of Households

Crop Complex	Percentage
Sweet potato	78
Taro and taro kongkong	15
Cash crops	05
Yam and winged bean	01
Mixed household gardens	01
Sugar cane	< 01

For the villagers, sweet potato is the staff of life. It is the main component of their diet, being consumed just about every day, and the major ration fed to pigs. One Kapanaran reflected on the overwhelming importance of the crop, noting "Just as a car will not run without petrol in the tank, so we cannot work without sweet potato in our stomach."

The first step in sweet potato cultivation is to choose cuttings of those varieties known to produce well in a particular environmental zone and plant them in mounds.[3] People harvest sweet potato from the mounds five to seven months after planting, the maturation time being influenced by the environmental zone. The method of harvesting and replanting is similar in all areas. Gardeners harvest from the sweet potato mounds for approximately two to three months. They do not take all the tubers in one mound at the same time, because doing so would deleteriously affect the growth of the trailing vines, which subsequently bear tubers in the soil between the mounds. After there are no more tubers in the mounds and a few months elapse, people harvest from the

compass and tape all the gardens of the sample of households (see Appendix B) between August 25 and October 9, 1976. Table 5.2 notes only the dominant crop or crops in the garden, not the numerous subsidiary crops. Because of the time of the survey it was impossible to measure the full extent of yam and winged bean, as they are usually planted between September and November and the harvest was completed before the survey period. Nevertheless, the area planted with yam and winged bean is estimated to be never greater than 10 percent of the total garden area. Gardens dominated by either sugar cane or leafy greens were more common in the past; although I did see some leafy green gardens in the village, none of the households in the sample had any. Both sugar cane and leafy greens are usually subsidiary crops in other types of gardens.

[3] Villagers plant at least 29 different varieties of sweet potato.

TABLE 5.3

Order of Harvesting Crops: Sweet Potato Gardens

1. Edible leaves of *Brassica* sp.
2. Edible leaves of bean plants
   (*Phaseolus vulgaris*), *Amaranthus tricolor*, and
   Brassica juncea.
3. Cucumber (*Cucumis* sp.).
4. Seeds of the bean plants.
5. Maize (*Zea mays*).
6. The leaves of *Rungia klossii* and the edible heart of the stem of
   Setaria palmifolia.
7. Sweet potato and pumpkin (*Cucurbita* sp.).
8. Edible leaves of *Abelmoschus manihot*.
9. Sugar cane.
10. Banana (*Musa* sp.).

trailing vines for several more months until most of the sweet potato has been taken. By harvesting from the vines lying between the mounds, villagers can obtain two crops from the same ground with one planting. Maize, beans, and leafy greens, which have been planted between the mounds, are first harvested, and the sweet potato vines bear tubers in the same ground after these earlier ripening crops have been consumed. When continued attempts to find more tubers are no longer fruitful, villagers rip the vines from their roots and completely till the soil, harvesting additional tubers not found before. When the vines are sufficiently dry, they are burned in the garden, and the area is planted again. This sequence is repeated in all successive replantings, though when the garden is ready to be fallowed, villagers do not till the soil to find the remaining tubers. Most sweet potato gardens have a life of from 12 to 18 months, though some people wait longer than 18 months to replant. Variations in the cropping period are related to the environmental zone in which the garden is located and to the number of times it has been previously replanted.

To understand the impact of commodity production and other external forces on the evolving subsistence system, classification according to the type of work (*kai'a*) in different environmental zones is essential (see Table 5.4).[4] This discrimination permits

[4] Although time periods are specified for fallowing in Table 5.4, neat, unvarying cycles of cultivation and fallow do not exist, and a few exceptions to these figures

the examination to be based on the "agronomic considerations of farming technique and management" (Brookfield 1968b: 417), which are more important in understanding system dynamics than crop combinations alone.

The following detailed discussion of the different types of gardening will illuminate three fundamental features of the subsistence system: (1) diversity in forms of agriculture helps alleviate the problem of seasonal rainfall; (2) labor inputs into subsistence production are not necessarily very seasonal because

TABLE 5.4

Types of Subsistence Gardening

Type of Work	Environmental Zone	Major Crops	Number of Times Replanted	Length of Fallow (Years)
Nanda kai'a	Inside rain forest	Taro, taro kongkong	0	14 +
Kaati kai'a	Forest fringe, near grassland	Sweet potato	1	9-20 +
Oba kai'a	*Miscanthus*-dominated community	Sweet potato	1-2	?
Ukau kai'a & *Baeha kai'a*	Short grasslands, dry soils	Sweet potato, yam, winged bean	1-3	5 +
Ha'i kai'a	Very steep, short grassland hillsides on sandy clay soil	Taro, yam	0	?
Hora kai'a	Moist soils of valley bottom	Sweet potato, yam, winged bean	3-4	8-15 +
		Taro, yam	0	4-7
Aruka kai'a	Swamp soils of valley bottom, dominated by *Phragmites*	Sweet potato, yam	3-4	15 +
		Taro, taro kongkong	0	4-7
Ana kai'a	Streambanks and rapidly draining soils of valley bottom	Sweet potato, yam, winged bean	2-3	15 +

can occasionally be found. In some cases, villagers prepared gardens on land that, according to them, had not been cultivated within memory.

different types of gardening vary considerably in the timing and intensity of labor requirements; and (3) subsistence affluence is based in part on full utilization of the range of options in subsistence production.

Nanda kai'a, or "forest-work," refers to the preparation of gardens within the rain forest as opposed to forest gardens bordering the grassland fringe. Villagers consider vegetation and soil characteristics in selecting garden sites. They prefer friable soil, which is detected by the ease with which young saplings can be pulled from the ground. Men claim that the presence of certain tree species such as *Castanopsis acuminatissima*, *Lithocarpus rufovillosus*, and *Garcinia* sp., as well as sprawling bamboo, indicates that the site is suitable for gardening. However, beliefs concerning which species is the best indicator vary; for example, some prefer sites with *Albizia fulva*, a nitrogen-fixing species, whereas others avoid sites with large *Albizia* trees because of the tree's reputation for causing temporary confusion and other mental maladies. Sites with a large vine called *maranda*, which has a reddish sap likened to blood, are generally avoided. Villagers believe that if the vine is cut, the spirit (*mara-ura*) of the vine will cause sickness or death to one of the family members of the person who cut it.

In *nanda kai'a*, cultivators first cut the sprawling bamboo six to twelve months before felling trees to allow sufficient time for the bamboo to dry. When the bamboo is dried, men and women using a bushknife clear the undergrowth, and men fell some of the large trees and defoliate others by ringbarking, firing the tree base, or pollarding. Especially large trees in the garden center are left standing because if felled, they would cover a considerable portion of the site. Gardeners allow a few trees to retain part of their leaf cover, thus facilitating the growth of taro, a plant requiring moist, shady conditions (Powell 1976: 121). After the task of clearing is finished, villagers allow the material to dry for two to three months before firing, with several burnings necessary before planting. Large, felled trees not fully consumed by the fire are left lying in the garden site.

The major crops in these forest gardens are taro and taro kongkong along with lesser amounts of banana, maize, beans, tobacco, cucumber, gourds (*Lagenaria* sp.), various leafy greens, pumpkin, and *Setaria palmifolia*. Villagers plant most of these gardens between September and December to take advantage of the increased rainfall during the early period of crop growth. They till the soil only where the crops are grown and, unlike in the grassland zones, do not construct drainage ditches. Gardens are planted only once before being fallowed for 14 years or longer.

Kaati kai'a, the creation of gardens inside the forest close to the forest-grassland boundary, differs slightly from *nanda kai'a*. Near the grassland boundary, soils are less friable and have a thinner layer of humus-rich material than those deep inside the forest. Clearing practices are similar to *nanda kai'a*, except that fewer trees are allowed to retain their leaves because the sweet potato grown here requires more sunny conditions than taro. *Kaati kai'a* resembles *nanda kai'a* in the method of tillage and in the absence of drainage ditches.

Villagers plant these gardens throughout the year except at the beginning of the dry season and in the midst of the rainy season, though if a short dry spell occurs in the rainy season, gardeners sometimes take advantage of the weather to burn and plant. However, most people plant gardens in the forest fringe from September to December. They do not cultivate the entire site at once; men divide the area by placing logs in a downslope direction and plant the sections at different times to ensure a continual supply of a variety of produce.

Sweet potato is the dominant crop in *kaati kai'a*. Villagers also grow yam, beans, maize, cucumber, leafy greens, *Setaria palmifolia*, pumpkin, sugar cane, and tobacco. To plant sweet potato, they first use an axe to remove tree roots at the mound sites. On hillsides, the soil is then heaped into convex-shaped mounds by digging out a slightly oblong portion of the hillside, with more soil being removed from the upper-slope portion, thus leaving a crescent-shaped scar just above the mound about 20 centimeters in height.[5] Sweet potato mounds are, on average, 8 centimeters high and 52 centimeters in diameter. Villagers state that these mounds are smaller than those found in the grasslands because the existence of tree roots makes it difficult to dig far into the soil and larger mounds would be more subject to erosion on the steep hillsides. In relatively flat forested areas, they prepare slightly larger mounds. The sweet potato mounds are placed fairly far apart, with an average density of 0.52 mound per square meter to accommodate the long, trailing vines that develop. The seemingly

---

[5] All data noted on the size of sweet potato mounds, planting density, and drainage ditches represent averages of many measurements, unless otherwise indicated.

haphazard spatial arrangement of the mounds in this zone reflects the presence of tree stumps and fallen logs.

In *kaati kai'a*, villagers plant two crops of sweet potato before fallowing for nine to twenty or more years. They believe that yields from a third planting would be poor because the soil could not support a third crop and the problem of weeding would be too difficult. By the time the garden is fallowed, secondary species are already encroaching onto the site, especially those that are purposely not weeded. To create future sites at which they can hunt certain phalangers, gardeners do not remove saplings of trees such as *Saurauia congestiflora* and *Ficus calopilina* because the animals will later feed in them during the fallow period.

If garden areas originally covered by forest are not fallowed for a sufficient time, they will eventually be dominated by the tall cane grass *Miscanthus*. Clearing and preparing gardens in this vegetative community is called *oba kai'a* or *"Miscanthus* work." Both men and women cut the tall grasses, allowing sufficient time for the vegetation to dry before burning it. The few, scattered trees in this zone are usually pollarded and later cut for firewood. Men dig small drainage ditches approximately 30 centimeters wide and 20 centimeters deep around the perimeter of the cultivated site and more rarely within the garden area. As in *kaati kai'a*, gardeners till the soil on the first planting only where the crops are placed, and sweet potato is the dominant crop. The planting schedule is also the same. However, the range of subsidiary vegetables grown here is commonly reduced to such crops as maize, beans, pumpkins, sugar cane, and only a few leafy green vegetables. In contrast to cultivation practices in the forest fringe, sweet potato mounds here are arranged in somewhat more orderly rows, and the mounds are more clearly convex in shape. People replant the garden once or twice, depending on how well the second crop of tubers produces.

Villagers also cultivate in the short grasslands. They use the terms *ukau kai'a* and *baeha kai'a* to denote the creation of gardens on relatively dry sites dominated by short grasslands. The term *ukau* refers to "grass" in general and *Themeda* in particular. *Baeha* means "strong," referring to the difficulty of cultivating grassland clay soil when dry. Some people restrict the use of the former term to the preparation of gardens on steep, well-drained hillsides dominated by *Themeda* and the latter to garden work on more gently inclined slopes dominated by a variety of grasses such as *Themeda*, *Imperata*, *Ischaemum barbatum*, and *Ophiuros tong-*

calingii. Methods of clearing and tillage are similar in both. Cultivating in the short, dry grasslands requires the least effort of all types of work, but gardens there also give the lowest yields of sweet potato per unit area.

The most preferred short grassland sites on relatively dry soil have a cover of either *Imperata* and the bracken fern *Pteridium aquilinum* or a cover in which *Imperata* and *Ischaemum barbatum* are abundant. Sites with the former combination are felt to be particularly suited to yam and winged bean as well as sweet potato, whereas those with the latter are considered good mainly for sweet potato cultivation. Because of their low fertility and dry condition, sites on steep hillsides dominated by *Themeda* are usually not cultivated except in two instances. First, if the area is relatively close to or adjacent to a residence, a gardener will use the site, because the negligible transport costs compensate for the lower yields. People enjoy the convenience of having a garden near their home to provide food when the demands of other activities make it impossible to travel to their more remote gardens. Second, if the *Themeda*-dominated site is next to a relatively flat, cultivated area with moist soils, a villager may also garden on the steep slopes. Utilizing two different, adjacent zones permits a variety of crops to be planted and different planting schedules to be used in the same garden area; for example, sweet potato can be planted on the dry hillsides more readily than in the poorly drained, valley bottom soils during the rainy season.

Drainage ditches are recurrent landscape features in short grassland gardens. The ditches are usually aligned in the direction of the slope and are designed to remove water rapidly to prevent waterlogging of the soil, which could produce landslides. The ditches, which surround the rectangular garden area and internally subdivide it into parallel strips, are placed five to eight meters apart and measure approximately 40 centimeters wide and 15 centimeters deep.

Compared with working in forest or *Miscanthus*-dominated communities, clearing in the short, dry grasslands is relatively easy. Women use a spade and bushknife to cut the grass, allowing the vegetation to dry before burning it. Then, after waiting one or two days, people till the entire garden area to a depth of 15 to 20 centimeters, preferably when the soil is slightly damp, which facilitates cultivation.

Complete tillage in grasslands performs several functions. It incorporates ashes and small pieces of dried, unburned organic material into the soil. It also breaks the capillary bonds to the soil surface, thus reducing soil water evaporation, a function especially important in the dry season. Clarke and Street (1967) observed that tillage markedly increases sweet potato yields in grassland areas in the Highlands because the improved drainage and aeration accompanying tillage increase nitrification and relieve the high reducing potential and toxic concentrations of iron and manganese characteristic of some untilled Highlands grassland soils.

Villagers plant various crop combinations in the short grasslands. If they grow yam and winged bean first, sweet potato will usually be the dominant crop on successive plantings. If sweet potato is the first crop, occasionally a small portion of the garden area is devoted to peanuts or winged bean in the second planting. Some people plant sweet potato as the only dominant crop throughout the life of the garden. The range of subsidiary crops is limited mostly to maize, beans, pineapple, *Setaria palmifolia*, and only a few leafy greens, as most crops do not do well here because the soil dries rapidly. Villagers usually plant sweet potato gardens on the steeper, well-drained hillsides from October to December,[6] but on relatively less-inclined slopes with dry soils they do so throughout the year, except when the soil is waterlogged or at the beginning of the dry season. The time for planting winged bean and yam is generally from September to November.

Villagers feel that it is necessary to wait a few days after the soil is tilled before planting sweet potato to allow the greasy condition of the exposed soil to dry. They build convex mounds averaging 17 centimeters high and 76 centimeters in diameter in somewhat aligned rows with a high planting density of 0.74 mound per square meter. People leave only a small space between mounds and prepare the mounds partially aligned in rows because the trailing vines, which grow poorly here, require little space.

Sweet potato matures here as early as five and one-half months after planting. Because of poor productivity, tubers carried by the trailing vines are harvested for a shorter period than in other zones. Depending on the distance from the cultivated site to the hamlet, villagers replant these gardens one to three times, with the greater frequency of planting possible near their houses be-

---

[6] Planting at the beginning of the rainy season enables crops to take advantage of flushes of nitrogen and phosphorus produced when the first rains fall (Nye and Greenland 1960: 110-111).

cause they throw refuse, ashes, and coffee pulp onto the gardens to increase fertility. With each successive planting, a larger percentage of the harvest is fed to pigs because the average size of the tubers produced decreases and the extent of the damage to tubers by the sweet potato weevil (Cylas formicarius) increases. Sometimes people replant these gardens with the sole intention of feeding the entire sweet potato crop to the pigs.

The length of the fallow varies, depending in part on the relative location of the site. Gardens near the hamlets are fallowed for as little as five to ten years, whereas places farther away are fallowed longer. Occasionally, people fallow a site for one or two years between successive crops before abandoning the site for a longer period. Such short fallows are possible because of the ease of clearing.

Another type of garden work in the short grasslands, ha'i kai'a, was more common in the past. Found only in the western sector of the village territory where a particular type of sandy clay occurs, these gardens are located on the top of steep hillsides where the dominant vegetation is Imperata and the fern Pteridium. After clearing the vegetation but before tilling the soil, a small fence is constructed on the downslope end of the garden to prevent soil loss on the steep slope. These gardens are usually small, with a shallow drainage ditch around the perimeter of the site; the only such garden that I measured was 65 square meters in size, whereas individual gardens in other zones are as large as 2,000 square meters. In ha'i kai'a villagers usually plant from October to December, though some do so as early as June or July if there is sufficient rain. They interplant taro and yam along with Saccharum edule, Abelmoschus manihot, and gourds, and after the crops are harvested they fallow the site. Gardening in this zone is not particularly common today because in dry conditions yields can be low compared to gardens in other zones.

Hora kai'a ("wet work") and aruka kai'a refer to garden work on flat land in the very moist-to-swampy soils of the valley bottom. Where the soil is very moist and Ischaemum polystachyum, Leersia hexandra, Ischaemum barbatum, and small, scattered Phragmites are cleared, the term hora kai'a is used. Ischaemum polystachyum is highly regarded as a fallow cover because of the reputed ability of its root system to impart good tilth to the soil. In swamps, where villagers cut dense stands of the taller reed grass Phragmites, they refer to garden work as aruka kai'a. Hora kai'a and aruka kai'a are felt to be the most physically demanding types of work in terms of labor inputs but also the most productive in gross yield. The high yields result from an adequate water supply throughout the year and an initial high fertility.

Clearing grass in *hora kai'a* is more difficult than in the short, dry grasslands because the vegetative cover is much denser. Women do most of the clearing, although occasionally men will help. They cut the vegetation at the base with a bushknife and spade and use a long pole as a lever to lift the mass of grasses to facilitate further cutting. During the next few weeks, people occasionally turn the pile of cut vegetation to ensure thorough drying before burning. Some go through the clearing and burning stages twice, allowing the vegetation to regrow for ten to twelve months before cutting and burning again. Such repeated cutting and burning weakens the root system and reduces the problem of subsequent weed regrowth.

Hora kai'a requires a more intricate and larger drainage system than that found in the short, dry grasslands. Because the water table is naturally high in this zone, the crops will rot unless the water table is lowered sufficiently. Around the perimeter of the neat, rectangular gardens, men dig large ditches approximately 1.2 meters wide and 90 centimeters deep. Draining into these is a series of smaller ditches 60 centimeters wide and 50 centimeters deep, which are placed parallel every four to seven meters within the garden. Men heap soil from the drains onto the edge of the ditches and mold it to form a smooth surface, which is later used as a walkway. Villagers wait about a month after digging the ditches to ensure adequate drainage before tilling the entire area to a depth of 6 to 12 centimeters.

The major crops planted vary. In most cases, the garden will have sweet potato in each of the four or more possible plantings. The time from planting to final harvesting of tubers carried by the trailing vines varies from 13 to 18 months. Sometimes, villagers grow yam and winged bean or yam and taro first and sweet potato on successive plantings. Less frequently, they plant taro and some yam and then fallow the garden for four to seven years before growing the same foods again. If the major crop is sweet potato, gardeners plant at any time except during the height of the rainy season. However, they usually plant yam, taro, and winged bean from September to November.

The sweet potato mounds, 25 centimeters high and 80 centimeters in diameter, are larger than those in short, dry grasslands to provide more drainage and to accommodate the larger tubers produced in this zone. Villagers prepare the mounds in neatly aligned rows spaced farther apart than those in the short, dry grasslands because the vines here grow longer. Preparing orderly rows is the most efficient means of planting the maximum amount of sweet potato possible, given the desire to harvest from both the mounds and the trailing vines. The neat spacing of rows ensures a higher level of net photosynthesis because the likelihood of the long, trailing vines overshadowing each other is reduced. The planting density is a low 0.43 mound per square meter. Other crops grown are maize, beans, pumpkin, gourds, *Setaria palmifolia*, tobacco, sugar cane, banana, and many varieties of leafy green vegetables.

The fallow length varies depending on the distance between garden and homestead. Close to the hamlets, fallows are as short as eight to ten years, whereas in areas farther away, gardens are usually abandoned for 15 years or more. People maintain that gardens are fallowed not because of a lack of soil nutrients but because either weeding becomes too burdensome or the fences have rotted, allowing pigs to destroy the gardens.

Garden preparation in *aruka kai'a* is similar to that in *hora kai'a* except in a few details. Clearing *Phragmites* is easier but drainage construction more laborious because of the difficulty in digging through the mass of thick roots and the need to construct larger drains to lower the higher water table, which is often above the surface. Villagers sometimes begin digging the drains up to a year before planting because of the difficulty involved. The drainage network reaches its highest level of complexity in *aruka kai'a*. Small drainage ditches feed into a larger one around the perimeter of the garden, and water from the latter flows into one major ditch up to 2 meters wide and 1.3 meters deep. One reason people prefer to establish adjoining gardens in *aruka kai'a* is to share the inputs into constructing the drainage system. People allow the site to drain for two or three months before planting.

All the crop combinations and rotations noted for *hora kai'a* are applicable to *aruka kai'a*, except that villagers do not plant winged bean here but do grow taro kongkong with taro and yam. Sweet potato is also dominant here. Because of the large tubers grown and the need for better drainage, sweet potato mounds are biggest in *aruka kai'a*. Some are as large as 1.0 meter in diameter and 27 centimeters in height. To accommodate the extensive growth of the trailing vines, the planting density is a very low 0.18 mound per square meter. Because of the large space between rows of sweet potato mounds, villagers can plant a considerable quantity of leafy green vegetables, many of which grow best in this zone. Some gardeners feel that sweet potato yields are better from the second planting than from the first. Lower initial yields may possibly result from an increase in the carbon-nitrogen ratio caused by the decaying root mass, which stimulates an increase in soil micro-organisms that consume nitrogen.

The planting schedule and the number of times a garden is replanted are similar to those noted for *hora kai'a*. Fallows are usually longer than 15 years, and villagers report that some sites in the swamplands have not been cultivated within memory because of fears concerning this zone (see below).

Ana kai'a, the third major form of work in the valley bottom, refers to cultivating on *ana bata*, a rapidly draining clay soil that occurs mostly along streambanks[7] and to a lesser extent farther away from the streams. The vegetative cover is composed of tall grasses such as *Phragmites*, *Miscanthus floridulus*, and *Saccharum robustum*, and shorter ones such as *Ischaemum polystachyum* and *Imperata*. Many villagers prefer not to cultivate near streams because overflowing water occasionally destroys the gardens. However, some take the risk, believing that sweet potato produces best on *ana bata*.

Some cultivation practices differ from those in *hora kai'a* and *aruka kai'a*. Because the soil drains rapidly, villagers make relatively small drainage ditches only 47 centimeters deep and 25 centimeters wide, and on the initial planting of sweet potato, most till the soil only where the crops are grown.

Gardeners plant either yam and winged bean or sweet potato as the dominant crops on the first planting and sweet potato on the two or three replantings that are possible. Occasionally, a portion of the garden is devoted to peanuts, a crop that does not do well in other soils in the valley bottom that are poorly drained. Villagers plant maize, beans, cucumber, banana, sugar cane, and various leafy green vegetables as subsidiary crops, and fallows are usually longer than 15 years.

At any one time, a household normally has gardens in three or four or more environmental zones. Such a strategy of diversifi-

---

[7] *Nondau kai'a*, another form of streambank cultivation, is distinguished from *ana kai'a* because in the former, large trees are felled. Taro and yam are the principal crops here, which are planted only once before fallowing. *Nondau kai'a* is largely restricted to the bank of one stream in Kapanara. No one in the sample and few in the village prepared such gardens.

cation permits the cultivation of a wide range of crops. It also provides insurance against the effects of unseasonal weather because a crop will have different degrees of susceptibility to rainfall variations depending on the zone in which it is planted.[8] Furthermore, because gardens in the numerous zones can be planted at various times and require differing intensities of labor expenditure, inputs into all phases of gardening can occur throughout most of the year. Table 5.5 lists the percentage of land cultivated by the sample of households in each environmental zone.

The villagers' cultivation practices in the diverse environmental zones have developed over a long period. Variations in the timing of planting, the construction of drainage networks, the crops grown, the method of tillage, and the preparation and spacing of mounds reflect an intimate knowledge of the potentials and limitations of the natural environment. Except in infrequent cases of extremely unfavorable rainfall conditions—which can be either too little or too much precipitation—the agricultural sys-

TABLE 5.5

Percentage of Land Cultivated in Different Environmental Zones, Sample of Households, 1976

Environmental Zone	Type of Work	Percentage
Moist soils of valley bottom	Hora kai'a	26
Short, dry grasslands	Baeha kai'a and ukau kai'a	18
Miscanthus-dominated community	Oba kai'a	17
Inside forest	Nanda kai'a	16
Forest fringe	Kaati kai'a	13
Swamp soils with Phragmites	Aruka kai'a	8
Streambanks and rapidly draining soils of the valley bottom	Ana kai'a	2
Very steep, short grassland hillsides on sandy clay soil	Ha'i kai'a	0

[8] The importance of preparing gardens in the moist-to-swampy soils—hora kai'a and aruka kai'a—cannot be overstressed in relation to both the normal dry season and the more extreme cases of drought. For example, during the Highlands-wide drought of 1972, when many communities experienced food shortages, Kapanaran valley bottom gardens produced adequately even though their other gardens in the drier hillsides and forest fringe suffered. Kapanarans supplied cuttings of sweet potato vines to needy areas.

tem has, since pacification, produced adequate and reliable yields enabling Kapanarans to satisfy fully household and ceremonial needs and even to sell a small surplus.

THE CHANGING AGRICULTURAL SYSTEM

Many studies attempting to document the negative impact of colonialism and the commercial economy on Third World rural communities present a dichotomy between "past" and "present," the "past" being represented by either the precolonial era or the period before incorporation into the global commercial economy. Although not all these researchers portray the past as a time of agricultural abundance (e.g. Shenton and Watts 1979; Bryceson 1980), they generally agree that in the "present," local food production systems have been undermined seriously (e.g. Kjekshus 1977; Nietschmann 1979; Elwert and Wong 1980). As Bryceson (1980: 282) observed:

> The underlying theme becomes one of colonial and neo-colonial violation of an internal balance and well being. . . . Pre-colonial peasant societies come to represent a 'golden age'; meanwhile the present is explicitly posited as a negation of the past.

Similarly, in much of the literature concerning the Melanesian culture region, traditional subsistence systems are characterized as being flexible and diversified, well adapted to the natural environment, and providing adequate, sometimes abundant, returns. These beliefs about precolonial agriculture are perhaps best exemplified in Fisk's concept of "subsistence affluence." Many other researchers have approvingly used the concept or similar ones in discussing traditional subsistence agriculture in the region (e.g. Lockwood 1971; Clarke 1973; Moulik 1973; Bonnemaison 1978; Connell 1978; Keesing 1978; cf. Young 1971).

I would like to challenge the conventional wisdom and argue that at least for part of the Eastern Highlands of Papua New Guinea, this dichotomy is too simplistic, masking important historical changes. My characterization of the subsistence system— flexible and diversified, well adapted to environmental constraints, and providing adequate, sometimes abundant, yields— applies specifically to the post-contact period. Thus, "subsistence affluence"—a condition now being undermined by the commercial economy—has not been an enduring feature of Kapanaran history, nor, most likely, of many other Eastern Highlands communities. Certainly, many Kapanarans themselves would be skeptical about applying the concept to precolonial conditions. The traditional tool kit, warfare, and beliefs all combined to limit the level of food output, with warfare being the major obstacle.

Compared with the steel tools of today, the precontact stone, wood, and bamboo implements were less efficient.[9] Men used stone adzes to fell and pollard trees, make fence posts, and clear the base of the tall cane grass *Miscanthus*. They dug trenches with wooden, paddle-like spatulas, and women tilled the moist-to-swampy soils with a smaller paddle. Villagers also employed a digging stick with a sharpened edge to perform a variety of tasks, such as tilling the better-drained, more compact soils. They cleared grass with both sharpened wood and bamboo cutters, and on the drier slopes where the vegetative cover is thinner, women also removed grass by hand.

Warfare, always a potential threat to the subsistence system, was endemic. Men attempted to kill their enemy in pitched battles; in small, stealthful raids; and in ambush. Men, women, and children were all potential targets, though killing a leading enemy warrior was a prime goal. Warfare was a major preoccupation, and, as Langness (1972: 174) noted for the Bena Bena of the Eastern Highlands, "The ideal [was] to be brave and strong to have a name—that is, to be widely known and respected, primarily as a warrior." Although it is impossible to reconstruct the exact frequency of intervillage warfare, several sources are suggestive. Older Kapanaran men, who once engaged in warfare, recount that peace between enemy villages lasted at most several months. Watson (1967: 71) argued that for another group of Tairora in the precontact period

> We may fairly assume . . . long periods of military quiet were few. Fighting may not have been 'constant' as it is sometimes described, if by that is meant a weekly affair. There can seldom have been whole seasons, however, let alone years, without interterritorial skirmishes or killings, or larger scale hostilities.

The constant threat of warfare also influenced the location of the stockaded hamlets. Villagers usually built their settlements close to the grassland-forest boundary for several reasons. The distance required to carry the planks to palisade the hamlets was

---

[9] For further information on precontact Highlands agricultural technology, see Watson (1967), Sillitoe (1979), Steensburg (1980), and Golson (1981a).

reduced. These locations were usually at a higher elevation than the rest of the valley, making it easier to spot the approaching enemy and making surprise attacks less likely. In addition, in case of defeat people were easily able to hide with their pigs in the forest (see also Newman 1965: 19).

The maintenance of conditions approaching subsistence affluence depended on the relative success of a group in warfare, but older Kapanarans recall that, although they were often victorious, they suffered defeat many times as well. When unsuccessful in fighting, their enemies burned their houses, destroyed their garden crops and fences, and killed and stole their pigs—all contributing to food shortages.[10] Indeed, the purpose of destroying an enemy's food supply was to weaken them, reducing their retaliatory capacity. Langness (1968: 193) described accounts of warfare before pacification among the Bena Bena: "The theme of fighting the strong is common. They say they lived well only at certain times—when Korofeigu was strong and other places were small."

The pressures of warfare detracted from the time available for subsistence production. Watson and Watson (1972: 576) reported the response of one informant in a nearby Tairora village to a question concerning the time spent on preparing for warfare:

We talked in the men's house, discussed preparations, alliances, who would help. Built palisade. Made emergency exit from men's house. We spent a great deal of time on all of this. We spent more time on fighting and talking about it than anything else. The women cried because they could not get enough help with the garden work (the men's part). Garden work suffered

[10] One indication of the impact of warfare and the relative degree of success in battle is the number of settlement site movements related to fighting. Only once or twice in this century, sometime between 1900 and 1928, were Kapanarans forced to flee their territory and seek refuge in friendly villages, though they later returned. Movements within their territory were more frequent. Older Kapanarans recall that between 1928, the time the first Europeans visited the area, and 1945, when an administration policeman settled the Kapanarans into one large village, they abandoned hamlet sites to move to another part of their territory from five to seven times—about once every three years. (Because villagers often lived in scattered hamlets, the number of movements that one person reports can differ from those of another.) Most of, if not all, the movements were related to warfare and destruction inflicted by the enemy, and it is highly likely that major damage to their gardens occurred when they were forced to move. Furthermore, there were often times when their enemies were able to damage Kapanaran gardens, but Kapanarans did not change residence.

because of men's preoccupation with fighting. Now the gardens are much bigger.

When fighting subsided, but the likelihood of attack was still great, men—armed with shield, bow, and arrow—would accompany women to the garden to protect them while they harvested. In some instances they were also able to work in the gardens, but at other times they thought it more essential to be on guard for a surprise enemy attack. In response to the need for security, villagers often grouped their gardens together.[11]

Although villagers planted gardens in various parts of their territory, they relied largely on ones close to the stockaded hamlets during fighting and only ventured to more remote ones when hostilities declined. As one elderly woman recalled, "We think of our enemy and plant nearby [the hamlets]." The vulnerability of more remote gardens, particularly those on the valley floor, was observed by an administration patrol officer who was visiting the better-drained grasslands had definite drawbacks. Planting a garden there early in the dry season was, and still is, risky He reported that after the Kapanarans made a surprise raid, their enemy fled, preparing new gardens and building new hamlets in naturally protected places in the hills; Kapanarans, unable to overrun the new hamlets, "had to content themselves with despoiling gardens lying on the floor of the valley" (PNG, Linsley 1948/49: 8).

The threat of warfare and the fear of hidden enemy sorcerers made people reluctant to plant gardens too far into the forest, where ambush was always a possibility. Old men also assert that they did not like to prepare gardens deep in the forest because of the difficulty of felling large trees with their stone adzes.[12] They

[11] Clarke (1966: 354) observed the Tairora practice of many families gardening within a common fence, and he correctly argued that such a pattern reduces the labor and materials required for fencing. However, such an aggregated pattern in gardening can also be found in cultivated areas without fences and is partly a continuation of the precontact custom designed to provide security. Villagers gave several additional reasons for gardening together. When other villagers are present, sorcerers are less likely to attack. Thieves will be deterred from stealing food if they fear that others might see them. Having people working in the same area enables gardeners to have social gatherings during periods of rest. Sometimes a villager will plant a garden alone at first, but when others see that his crops have grown especially well, they will come and prepare gardens nearby.

[12] Although it is likely that men using stone adzes pollarded more trees and felled fewer ones compared to today—a practice that would have reduced the labor

preferred instead to work in the smaller secondary forest near the grassland where the view of the surrounding countryside was better.

Before pacification, gardening in the valley bottom was not as widespread as it is today because villagers working in such low-lying areas were more susceptible to surprise enemy attack. When cultivating there, villagers preferred to work where their vision was not restricted, usually avoiding small pockets confined by sharply inclined hills. Within the low-lying areas, *hora kai'a* was more extensive than *aruka kai'a*, the latter usually being limited to the edges of the *Phragmites* swamps. Given the precontact technology of wooden spatulas for digging drainage ditches and sharpened bamboo and wood for clearing grass, the task of creating extensive garden areas with an intricate drainage system in the swamps was considerable.[13]

Their beliefs also constrained use of the swamps. Villagers feared that many of the larger areas covered by *Phragmites*, especially where the soil was not firm enough to support the weight of a man, were inhabited by worms or *ahe'u* that reportedly performed lethal sorcery on family members working in swamp gardens. These worms supposedly collect any human refuse, such as feces, urine, and hair, or pieces of leftover food dropped by individuals and use these in rituals that can result in the death of either the person leaving the material or one of the family members. As children are careless, as well as the weakest family members, they are the ones most likely to be injured. Those who did cultivate at the edge of the swamps sometimes performed special rituals designed to frighten away *ahe'u*. They also burned pig grease in the garden area hoping that the resulting smell would drive out the worms. Occasionally, gardeners abandoned swamp sites when bad omens occurred; the collapse of a drainage ditch wall in the swamps and dreaming about the death or sickness of a family member were considered indications that continued cultivation at the site would result in harm to a household member.

---

inputs necessary for clearing (e.g. Clarke 1971; Sillitoe 1979)—the task would still have been demanding physically with the dull stone tools.

[13] Schindler (1952), in an early study of the agricultural system of the Gadsup people next to the Aiyura Agricultural Experiment Station, also noted that in the precontact era villagers did not cultivate extensively in the swampy flats because of the difficulty of draining areas covered by *Phragmites*. However, various environmental and demographic factors did lead to swamp cultivation in the Western Highlands (see Golson 1981b).

These beliefs are still current, especially among older villagers, and account for the fact that much of the swampland cultivated in 1976 was not gardened before within memory.

Before pacification, gardens in the well-drained grasslands (*baeha kai'a*) performed an especially crucial role. When many gardens throughout their territory were destroyed during fighting, villagers would first prepare new ones in this zone because establishing gardens there required the least time compared to other areas; there was no need to wait a month or more for ditches to drain the site (as in the low-lying areas) or several months for vegetation to dry (as in the forest fringe). However, reliance on the better-drained grasslands had definite drawbacks. Planting a garden there early in the dry season was, and still is, risky because the lack of soil moisture makes it difficult for planting material to thrive. Even after successful establishment, crops are still highly susceptible to periodic dry spells. In addition, the number of subsidiary crops planted, particularly leafy greens, is much less in *baeha kai'a* than in most other types of gardening.

In response to the constraints outlined, villagers grew most of their sweet potato in the forest fringe; in the short, dry grasslands; and to a lesser extent in the valley bottom. Cultivation in low-lying areas increased during relatively peaceful periods. A considerable amount of the taro and yam production was from *ha'i kai'a* in the short grasslands as well as from the forest and low-lying areas.

Subsistence affluence was thus not a constant, enduring characteristic of the production system during the precontact era. Conditions oscillated.[14] At times, warfare and hunger plagued the villagers. However, when peaceful conditions prevailed or when Kapanarans were victorious in battle, subsistence production was more secure and the pig herd expanded; at such times, people recall, they lived quite well and the food supply was abundant.

Pacification of the area in the late 1940s contributed significantly to a more stable food supply. The benefits of warfare cessation were clearly evident in the contrasting conditions of those who had ceased fighting and those who were not yet under administration control. For example, the administration patrol officer who visited the area in 1949 reported that most villages had stopped fighting and that there "is no shortage of food in any area and in

---

[14] Fisk, who coined the phrase "subsistence affluence," never suggested that it applied to all regions in Melanesia (see Young 1971; Lacey 1981).

most a surplus exists." However, Kapanara and their enemy continued to raid each other over a five-month period, and consequently the two were "not as well off for food as the others, principally due to the unsettled conditions prevailing there" (PNG, Linsley 1948/49: 12). Berndt (1953: 118), one of the first anthropologists to conduct research in the Highlands, made similar observations in comparing the situation before and after the imposition of colonial control in the Eastern Highlands:

> Whereas in former times gardens were abandoned or destroyed at intervals during the fighting (with ensuing periods of scarcity and hunger until new gardens commenced to bear), cultivation can now proceed relatively undisturbed.

Villagers could not significantly expand agriculture in the valley bottom, the zone most productive for gardening, until after pacification. Gardens in the low-lying areas are especially important in relation to the marked dry season, because production here is less affected by a shortage of rain than in other zones. An increase in the cultivation of valley bottom sites accompanying the cessation of warfare has been noted for other Eastern Highlands groups as well (e.g. Salisbury 1964; Newman 1965). As Salisbury (1964: 4) observed: "Peace, in short, permits land use to be rational, to be based on the potentialities of the land, rather than on the likelihood of being raided."

By the time the area was pacified, villagers had already obtained steel tools. In 1928, two German missionaries, the first Europeans to visit Kapanara, gave the villagers their first steel tools, an axe and a knife. Additional axes were obtained through trade with other villages, and by the mid-1930s steel axes had replaced stone adzes. A few years later, the bushknife became the major tool for grass clearing, but it was not until the late 1940s that the use of the steel spade became widespread.

Steel tools are superior in cutting ability to the duller stone adzes, wooden spatulas, digging sticks, and bamboo cutters they replaced. However, researchers have differing views on the resulting benefits. Many assert that the introduction of steel tools significantly reduced labor inputs into gardening. For example, Salisbury (1962: 220) estimated that villagers could fell trees and make garden fences three to four times faster with steel tools, figures supported by several field demonstrations involving the use of stone and steel (see Townsend 1969: 204; Clarke 1971: 175; Steensberg 1980: 38). In contrast, Sillitoe (1979) asserts that

the time savings may have been overestimated. Comparing effi-
ciencies in felling and pollarding trees, making fence posts, and
removing roots and stumps in forest and *Miscanthus* communities, he reports that steel tools are, on average, only 1.5 times
faster (with a range from 1.04 to 2.2). He correctly maintains that
time alone, however, is an inadequate measure of comparative
utility. His observation that working with stone tools is physically harder, more jarring, and painful is corroborated by the rec-
ollections of older Kapanarans. Also, because steel tools are more
effective cutting instruments, people perform additional gardening tasks not undertaken previously with stone implements, thus
lessening the comparative time efficiency of steel but creating
the potential for increased agricultural yields; for example, when
using steel, the villagers he observed in the Southern Highlands
Province cleared more tree roots and stumps, making room for
more crops to grow, and they cut up the vegetation into smaller
bits, facilitating more complete burning, which improves fertility.

These studies have emphasized the role of steel in cutting trees
and, to a lesser extent, *Miscanthus* cane. However, Kapanarans,
like other Eastern Highlanders, also prepare many gardens in several grassland zones. The use of the bushknife for clearing grass
and the spade for tilling soil and digging drainage ditches resulted
in considerable time savings (see also Golson 1981a; Steensberg
1980). Sorenson (1972: 369) reported that the Fore, another Eastern
Highlands group to the west of Kapanara, traditionally preferred
to make gardens within the forest zone, but with the introduction
of the spade, which made the task of clearing grass and cutting
through the sod easier, the Fore increased gardening activity in
the grasslands.

The cessation of warfare and the introduction of steel technology enabled villagers to increase the extent of gardening in the
valley bottom, though still placing more emphasis on *hora kai'a*
than on *aruka kai'a*. They did not significantly expand *aruka kai'a*
until the 1970s because the fear of *ahe'u* remained strong. People
increased the frequency of *nanda kai'a* and planted farther inside
the forest. Taro production was gradually transferred to *hora kai'a*
and *nanda kai'a* as less emphasis was placed on *ha'i kai'a*, because
forest and valley bottom areas are less subject to dry spells and
produce better taro yields. Increasing the extent of gardens in the
valley bottom and inside the forest was a function of the general
willingness to cultivate farther away from the hamlets as the fear
of enemy attack subsided.

The introduction of steel technology and, more importantly, the cessation of warfare ushered in a period of subsistence affluence. Agricultural yields increased, and garden output became more reliable. The entire range of environmental zones could now be exploited, providing a more substantial buffer against the vagaries of the weather. In addition, the pig herd was now safe from periodic decimation by the enemy, and, according to some informants, the herd size increased in the 1950s after pacification (see also Sorenson 1972: 369).[15] Boyd (1981: 92) recorded similar changes among the Ilakia Awa to the southwest: "With the decline of warfare, Ilakians say they increased the size of gardens to provide a larger and more stable supply of food for people and pigs, cultivated virgin land much farther away from the village, and increased the size of their pig herd."

Along with pacification and steel technology, the introduction of new food crops and new varieties of precontact species in the last fifty years has also benefited the agricultural system. Two crops that are now a major part of the diet—corn and taro kong-kong—reached Kapanara in the 1920s and 1930s respectively (possibly the early 1940s for the latter). Subsequent to pacification, Europeans introduced numerous, current subsidiary food crops such as peanuts, peas, tomatoes, shallots, Irish potatoes, various beans, several varieties of banana, and greens such as lettuce and cabbage. In addition, as intervillage warfare declined, the rate of introduction of new varieties of precontact crops—particularly sweet potato—through traditional trade routes increased dramatically.

Another change that was eventually to have far-reaching consequences for subsistence production was the introduction of commodity production. In the mid-1950s, the *luluai* of Kapanara obtained free coffee seedlings from the Highlands Agricultural Experiment Station at Aiyura and planted them in the village. After selling his first harvest a few years later, other villagers became eager to grow the crop, and with the assistance of agricultural extension officers they jointly cleared and planted a common plot. Within the plot, each man was responsible for caring for his own group of trees. As elsewhere in the Highlands, people soon abandoned the communal plot and planted individual hold-

[15] The pig herd, however, was occasionally reduced because of anthrax epidemics both before and after pacification.

ings (see Brown 1972: 89; Finney 1973: 62).[16] They slowly expanded their coffee holdings, spurred by the need for cash for domestic and ceremonial needs and by the introduction of taxation in 1958.

Coffee production and subsistence agriculture have always been integrally related. Coffee and subsistence gardens can compete for land because coffee is a relatively fixed feature of the landscape and food crops are not interplanted with mature coffee trees; coffee begins to bear about three years after planting, reaches maximum production four to five years later (PNG, Coffee Marketing Board 1972: 3), and yields for 25 to 30 years if properly maintained (Dwyer 1954: 4). Although villagers have planted coffee largely on prime agricultural land near the hamlets, the degree of displacement of subsistence agriculture has not been a particularly serious problem as the coffee groves are relatively small and agricultural land is abundant in the village. The entire area under coffee trees in 1977 was only approximately 0.2 square kilometer out of a total of 19.5 square kilometers of grassland within the village territory. Nevertheless, people planted a considerable amount of coffee during 1976 and 1977 in response to high coffee prices. For example, in a survey of the coffee holdings of the sample of 13 households in mid-1977, 29 percent of the coffee was three years old or younger, thus revealing a recent major expansion of coffee. Continued planting of coffee near the hamlets combined with the presence of the cattle projects will eventually result in a lack of available land for food gardening near the hamlets.

Villagers use a variety of sites to grow coffee. Sometimes it is planted on the sites of abandoned and destroyed houses where household refuse has increased the fertility of the area. Occasionally, people clear fallow vegetation specifically to grow coffee, and they interplant a few vegetables such as maize, beans, leafy greens, and pumpkins to efficiently use the space between the coffee seedlings. The most common method is to establish coffee in a food garden ready to be fallowed. Villagers often reduce the number of times they replant a plot with food crops if desiring to grow coffee. For example, a gardener could normally plant in

---

[16] In some parts of the Highlands, *luluais* maintained control over such communal plots when they were abandoned, giving them larger coffee holdings than others. In contrast, in Kapanara one of the poorer villagers, a mute, obtained the plot.

drained swamplands four times or more but may do so only twice before establishing coffee in the plot.

The cultivation of coffee has stimulated increased *aruka kai'a*, *hora kai'a*, and *ana kai'a*, because coffee grows best in the valley bottom. Coffee on dry hillsides is not as productive, though villagers still plant some on slopes in fallowed gardens near the hamlets. In the sample of 13 households, 87 percent of the coffee was on valley bottom soils and the other 13 percent on the dry, well-drained hillsides. The planting of coffee on the most fertile land is widespread throughout the Highlands.

The shift to more subsistence agricultural production in the valley bottom has been a trend continuing since pacification and accelerating after coffee was introduced. Gardening in *hora kai'a*, *ana kai'a*, and *aruka kai'a* is more intensive than in other types of garden work, whether measured in terms of the number of years a particular plot is cultivated over time or by the labor input per unit area over time. Such a voluntary intensification of garden labor is contrary to the assumptions in Boserup's (1965) model, in which she claims that people only intensify agricultural production if forced to by population pressure.[17] In Kapanara, villagers also prefer to cultivate in moist-to-swampy soils because production is less affected there by dry spells than in the other zones. Thus, both the desire to prepare suitable sites for coffee gardening and environmental constraints have made intensification a worthwhile alternative.

Although various forces have modified the system of subsistence agriculture, its primary purpose remains the satisfaction of

[17] Another important issue in examining Boserup's hypothesis is whether the more intensive gardens have a higher or lower productivity ratio than extensive gardens, but I do not have sufficient data on this. Gathering such data is too difficult a task in Kapanara. Production would have to be measured not only in different environmental zones but also for successive replantings in the same zone. Another problem is the representativeness of the data; several samples from different gardens within the same zone for each successive replanting would also have to be obtained. In addition, sweet potato production is highly variable. Allan Kimber (personal communication 1977) of the Highlands Agricultural Experiment Station at Aiyura noted that yields of sweet potato may vary substantially according to variety. Furthermore, in sweet potato yield trials using seven plots 38.4 square meters in size for each variety tested, the yields from one plot compared with another planted with the same variety sometimes differed by as much as 459 percent! With such variation, relatively large plots would be needed to ensure the representativeness of the data. The much more modest attempt that I made to measure yields was largely thwarted by the more determined efforts of the village pigs that raided my sample sites.

household consumption requirements. Another component of the subsistence system, pig husbandry, is more integrally related to the system of reciprocal exchange. No understanding of Highlands subsistence production is complete without an analysis of pig husbandry.

PIG HUSBANDRY

As in other Highlands areas, pigs play a central role in the lives of the Kapanarans. The possession and exchange of pigs are influential in attaining status and prestige. Almost every major event, such as birth, sickness, death, initiation, first menstruation, marriage, and the abrogation of food taboos, is accompanied by the killing of pigs and the distribution of pork. Hence, full participation in social life normally requires that pigs are owned. Pigs can be exchanged to fulfill obligations and create new ones. They are also highly valued as an important source of cash income. In a diet traditionally deficient in high-quality animal protein and fat, pork is a valuable component. Villagers also have a strong emotional attachment to their pigs. Of the 93 married households in Kapanara, 86 percent cared for at least one pig during three censuses conducted in July 1976, March 1977, and November 1977.[18] The pig per capita ratios for these three periods were 0.85:1, 0.48:1, and 0.67:1 respectively.

Within the village territory, the pig population is almost entirely domesticated; the few wild pigs that roam the forests were originally cared for by villagers. The current pig population is a cross between the native razorback pig and the larger European pig, as villagers had previously purchased a few European breed and razorback-European crossbreed pigs to improve their herd. Many feel that crossbreed pigs grow faster and are fatter than their traditional counterparts.

Pig husbandry, like agriculture, is a precontact tradition that has changed through time. During the period before colonial

[18] The censuses were based entirely on verbal information provided by household members. No attempt was made either to count or to tag the pigs myself. I collected data on the number, size, and sex of the household's pigs, and on ownership, agistment, births, deaths, prestations, sales, and purchases. I cross-checked household interviews with an almost daily record that I kept of pigs killed in the village as well as with information provided by others. Villagers' statements proved to be highly reliable. Unlike people in some other areas of the Highlands (see Feachem 1973: 28), Kapanarans do not attempt to conceal the number of pigs they own.

administration, pigs usually slept in special compartments in the women's houses at night. If relatively peaceful intervillage relations prevailed, they foraged for themselves during the day; at other times, villagers led them on a leash and tethered them where they were working, so that in case of attack they could lead their pigs to the forest and hide them from the enemy.

With the cessation of warfare, the eventual disappearance of sexual residential segregation, and the influence of government officials and missionaries who warned of the health hazards of sleeping near pigs, villagers abandoned these practices, though occasionally a small pig can still be found sleeping in one of the houses. Other aspects of the system before contact, such as the belief in the efficacy of ritual both in promoting the good health and fecundity of one's own pigs and in killing or debilitating another's, are still quite prevalent.

The pig house, a roofed structure smaller and less sturdy than a villager's house, is also a traditional part of the system. It is an alternative residence for pigs; instead of returning to the hamlets after foraging, a pig goes to the pig house to sleep and be fed. The distance between pig house and settlement is sometimes considerable and in 1976 ranged up to two kilometers. Villagers list several reasons for having pig houses. For example, if a pig invades and destroys part of a garden, its owner can tether it or confine it at the pig house for a few days in the hope that the pig will "forget" about the garden and not raid it again. Some individuals prefer to perform rituals on their pigs in the secrecy of their pig houses. Particular types of pig magic must be performed in secret, because once discovered and used by members of other households, the magic loses its power for the original practitioner. Finally, and most importantly, we must consider the prohibition on men from eating certain pigs, a custom based on men's fear of pollution from their wives. Men worry that when their wives feed their pigs, the animals may also consume some of the women's bodily fluids, such as menstrual blood, that may inadvertently contaminate the feeding area. Men believe that if they consume such polluting material from their wives, it can induce infirmities such as blindness or even death.[19] However, the fear in relation to consuming pork is not of menstrual blood or of

[19] Most women will not eat pigs that they feed either, though this should not be regarded as a strict restriction for women because several women did eat pigs that they fed during the fieldwork period. This voluntary restriction by females may result from the bonds of affection for their pigs. Some state that pigs are like children because women give them food.

other female bodily substances in general, but only of those from one's own wife, because men eat pigs fed by other women. The act of sexual intercourse appears to have special significance in specifying which women's pigs cannot be eaten, because a few informants confided that they also could not eat pigs raised by a woman with whom they were having an illicit affair. The prohibition on a man from eating pigs raised by his wife is so strong that he will not eat the pig cared for by another woman if it is cooked in the same earth oven as his own because of the possibility that his pig will contaminate the other one. A man will also not eat his neighbors' pigs that are fed close to his house, as these animals may have wandered through his hamlet consuming not only scraps of discarded food but also his wife's discharged bodily substances that accidentally fell on the ground. Thus, if villagers want to give him pigs in an exchange, they must feed them in a location far from his residence, such as in a pig house elsewhere in the village territory.

The use of pig houses far from the hamlets in the era before colonial control depended partly on the state of intervillage relationships. If relatively peaceful relations prevailed, villagers were more likely to make the once or twice daily trip to feed their pigs at the pig houses. The scheduling of important ceremonies such as male initiation also affected the use of pig houses; at such occasions, many pigs were killed, and in order to be sure that others within the hamlet could eat the pigs, the animals were fed at pig houses far from the residences.

With the cessation of warfare and a general increase in the size of the pig herd in the early 1950s, villagers used pig houses more frequently. However, in the mid-1960s the number of people using them dropped dramatically concomitant with the first suspected case of sangguma, a Tok Pisin term for a form of assault sorcery believed to be performed by a group of men from another village who supposedly violently attack a lone victim by driving spikes into his body or by removing his internal organs (see Glick 1972). Unlike traditionally recognized illnesses caused by sorcery, no known cure for the surely fatal sangguma attack exists. Villagers assert that the first such attack was made by members of a nearby village in 1963 or 1964, and after the incident, many feared going alone to feed their pigs far from the hamlets. The danger of raising their animals in pig houses is compounded because in order to familiarize pigs with the house, some members of the household caring for them must sleep there for several weeks to ensure that the pigs will continue to return to the pig house. Alone in a far away pig house, an individual is a prime target for *sangguma*. Only 20 percent of the 80 households caring for pigs had pig houses. However, 56 percent of the pig houses were located within 50 meters of the residences, and thus pigs raised in them could not be eaten by other men of the hamlet.

Following tradition, the task of caring for pigs falls largely on the household women today. If conscientious, a woman feeds her pigs once in the morning and again in late afternoon when the animals return from foraging. Such a level of care is usually reserved for pigs residing at the hamlets. Pigs that sleep at pig houses or other areas far from the residences receive rations less often. Feeding is necessary to ensure that the animals return home and do not become wild, and also to partially satisfy their hunger and therefore lessen the likelihood of food gardens being destroyed.

An owner often does not care for all his pigs. Agistment is common. Fifty-eight percent of those households raising pigs cared for at least one pig belonging to another household. One person had agisted pigs with seven different households. Agistment has several advantages for the owner. It reduces his own household's workload of feeding pigs. If sick or hungry for pork, the owner can request that the person caring for his animal kill and give it to him to eat, provided that the pig was not raised in his own hamlet. Furthermore, the chance of sickness spreading to all of one's pigs is reduced.

The person caring for the agisted pigs may also benefit. When a sow bears a litter, the owner usually gives him several piglets. If the pig is killed for some reason, the owner often rewards the person's labor expended in raising the pig with cash sums up to K30 or with a live piglet.

Agistment enables a man without any pigs to obtain a few piglets and to eventually build up his own holdings. Alternative means of obtaining them are also available. A few villagers purchase piglets from others in Kapanara or from members of other villages. Also, the elderly occasionally give pigs to their children, though upon the death of the parent, the son or daughter kills many of the large pigs received to satisfy mortuary obligations.

Although owning many pigs has never been as critical to aspiring big-men in Kapanara and the Tairora area as it has been in many other Highlands regions, it nevertheless aids a man in his pursuit of prestige. However, owning many pigs will not, in itself, guarantee prestige or a position of leadership. Also, several alternative means for achieving prestige, such as cash cropping, exist today, and some prominent village men possess few pigs. Nonetheless, pigs are still highly valued as items of exchange, and villagers unanimously agree that pig husbandry will remain a central focus of their economic and social life.

Pig husbandry in Kapanara has a markedly intravillage focus. To use Strathern's (1969) terminology, people employ a strategy of "home production" in which most pigs used in prestations are raised by the local group, in contrast to a strategy of "finance," which relies more heavily on utilizing links outside the local group to obtain pigs used in exchanges. Similarly, Kapanarans depend largely on natural increase as opposed to purchasing piglets from people in other villages to expand the herd size; 97 percent of the pigs added to the village herd during 1976 and 1977 were born in the village. Ninety-three percent of the pigs killed during this period were given to others in Kapanara, and 98 percent of the pigs owned by Kapanarans were raised in the village. This intravillage focus is partly related to the pattern of marrying people from the village, because many pigs killed or agisted are given to affines. It may also be a carry-over from the time before pacification when most surrounding villages were enemies and enduring relations with them concerning pig husbandry, such as agistment, may have been too risky.

The only significant link in the pig husbandry system with other villages is through the sale of live piglets.[20] Between July 1976 and November 1977, 23 percent of the Kapanaran piglets were sold to people in other villages. Eighty percent of those sold were sent to the south, with most of these destined for the Obura region (see Map 1.2). A further 14 percent were purchased by people in the Auyana region and in other Tairora villages to the southwest where many kin links were established before contact. The price received for a piglet in the mid-1970s was most commonly between K40 and K50 plus an additional payment of traditional trade items and valuables such as arrows, plumes, net bags, leaf mats, and grass and bark skirts. The trade is based on differences in genetic and environmental factors. Kapanaran pigs have a greater percentage of genetic inheritance from European breeds and are thus felt to be superior to those farther south. Also, the environment constrains pig production in the Obura region.

---

[20] Kapanarans rarely sold live, large pigs to people from other villages in 1976 and 1977.

The grasslands that occur there are mostly on steep, well-drained slopes; there are no extensive swamplands. Such a grassland environment does not provide good foraging. In contrast, Kapanara can support a larger pig herd because its swamps provide excellent foraging. The sale of live piglets to the Obura area has only become a significant source of village cash income in the 1970s.

The primary reason for raising large pigs is not to sell but to kill and give them to others on special occasions to satisfy debts, create new ones, and obtain prestige and cash. The nature of the prestations and counterprestations involved depends on the social ties linking the parties. An important series of exchanges concerns the *kandere* relationship.[21] An adult man or woman can kill and give a pig directly to his or her own older *kandere* or to the *kandere* of one of their children. The individual making the prestation also usually gives beer and raw and cooked foods. Normally, the pig owner will designate one to three individuals as the major recipients, who will distribute the pork, beer, and vegetables to others who are also considered *kandere*. These major recipients then pool money from those who received a portion of the prestation and present the cash to the pig owner. The actual sum given varies, depending largely on the size of the pig killed, but it is usually between K100 and K150. More rarely the recipients return a killed pig and a smaller amount of cash. Traditionally, villagers reciprocated with such items as bows, arrows, net bags, and shells, but these are no longer given in the exchanges. In other transactions, a man's "brothers" who are agnates and, to a lesser extent, those who are cognates are the major recipients. If a man kills a pig and distributes it to many of his "brothers," no specific obligation exists to repay the gift; feelings of solidarity were and still are strongest among fellow agnates, who ideally should be continually involved in such prestations and other forms of assistance. However, if one "brother" is specifically presented with a pig, he then has an obligation to reciprocate in the future by killing and returning a similar-sized pig. Numerous other social relationships are also involved in prestations.

Sometimes an owner's plans for giving a pig in a future exchange are disrupted when his animal is killed destroying a garden. Then he must decide quickly what to do with it. Most pigs are given in prestations. Some are marketed in the village, but the returns from marketing are poor, averaging about K14 per pig sold. Of the

[21] See footnote 11, Chapter Three.

289 pigs reported killed from June 20 to October 12, 1976 and from January 20 to December 13, 1977,[22] villagers cooked, cut, and sold by the piece 29. Usually only half the pig is marketed, and the rest is given to others such as neighbors and "brothers," with the expectation that the recipients will reciprocate with a piece of pork in the future.

Thus, the reasons for killing and exchanging pigs vary considerably. Table 5.6 lists the percentage of all pigs killed for various reasons for the periods June 20 to October 12, 1976 and January

TABLE 5.6

Reasons for Killing Pigs

Reason	Percentage
Spoil garden[a]	34
Affinal and *kandere* exchanges[b]	15
Other social reasons[c]	15
Mortuary practices[d]	10
Other[e]	10
Sickness and healing[f]	7
Drunk-related[g]	7
Reciprocity for labor[h]	2
n = 289	

[a] 'Spoil Garden' refers to pigs killed because they damaged gardens.

[b] 'Affinal and *Kandere* Exchanges' includes pigs given to affines on such occasions as visiting and the birth of a child. *Kandere* exchanges are described in the text.

[c] 'Other Social Reasons' refers to all social occasions not covered in the other categories, such as killing a pig for a party to celebrate a narrow escape from death, release from jail, or to reciprocate to those who helped in a fight.

[d] 'Mortuary Practices' refers to pigs killed for ceremonies concerning the death of an individual. Pigs are also given to reciprocate for the burial of the deceased and are killed when abrogating food taboos initiated after the death of a relative.

[e] 'Other' includes consumption of pigs to satisfy hunger, the killing of pigs that were sick or wild, or unknown reasons.

[f] 'Sickness and Healing' includes pigs slaughtered to feed a sick individual as well as the ritual healer and the guests who attend healing ceremonies.

[g] 'Drunk-Related' refers to the exchanging of pigs initiated because of an action while drunk, such as licking the face of one's *kandere*, crying with a brother, and hitting or scolding a 'sister' for not properly helping her 'brothers.'

[h] 'Reciprocity for Labor' includes the killing of pigs to reciprocate for labor provided by others for such tasks as garden preparation or housebuilding.

[22] During this period, I obtained reports daily or every few days on pigs killed in the village. I estimate that these verbal reports included 80 to 90 percent of the pigs that were killed or died.

20 to December 13, 1977. The table lists only the owner's original reason for killing the pig and not its ultimate destination.[23] The only category not fully represented is "reciprocity for labor," because many villagers built new houses between November and January and gave pork to those providing assistance, but this period is not covered in the table.

In approximately 14 percent of all prestations involving pigs, people reciprocated with cash. Considering the additional large number of piglets sold and the number of occasions pork was marketed, villagers understandably declare that pigs are *bisnis* just like coffee and cattle raising.

As well as the social and commercial functions of pig husbandry, the animals also play an important role in the subsistence system. The carbohydrates fed to pigs are converted into high-quality protein and fat, a process that is poor in terms of energetic efficiency but important in its contribution to the protein- and fat-poor diet (Rappaport 1968: 68). Vayda, Leeds, and Smith (1961: 71) have suggested that in the absence of means for storing tubers for long periods, feeding vegetable produce to pigs during years of good harvests is important in creating "food reserves on the hoof" to be consumed or traded in times of poor harvests. Pigs also aid communities by removing garbage and feces from hamlet areas and thus ultimately provide a link to the detritus food chain (Rappaport 1971: 128).

In some aspects, pigs can play a complementary role to the agricultural system. When ready to fallow gardens, villagers sometimes allow pigs to root for unharvested tubers, thus efficiently harvesting the remaining crop. Furthermore, the activity of pigs in the garden decreases soil bulk density and increases site fertility as pigs deposit feces and urine. Some Highlands groups such as the Maring also allow pigs to dig in gardens before replanting (Rappaport 1968: 57), but Kapanarans usually do not; the sweet potato crop in a garden here is often at different stages of maturity, with one area ready to be replanted and another still producing tubers, and the rooting of pigs would destroy the drainage ditch system. In addition, following suggestions of agricultural extension officers, a few villagers have built wire or wooden enclosures to confine pigs when they are to be fed or when they destroy gardens; after pigs have improved the soil condition, the owners cultivate the site.

[23] For example, some pigs killed while invading a garden were given to *kandere,* but in Table 5.6 the incident is recorded as a case of spoiling a garden, not as a *kandere* exchange.

Although pigs provide numerous benefits, they are also a burden. Various researchers have noted the large percentage of the daily harvest fed to pigs in the Highlands. Waddell (1972: 119), for example, found that 49.2 percent of the produce that a group of Raiapu Enga brought home was given to them; the pigs' daily ration of sweet potato was 1.4 kilograms per pig. The Tsembaga fed pigs 54 percent of the sweet potato and 82 percent of the cassava growing in their gardens (Rappaport 1971: 127). In Kapanara also, pigs consume a considerable portion of the agricultural production.[24] Pigs at the hamlets received a per capita average daily ration of 1.66 kilograms of whole, uncooked sweet potato and 0.24 kilogram of scraps consisting mostly of sweet potato peels.[25] The least any household provided was 0.5 kilogram per pig per day, and the most was 2.86 kilograms per pig per day for very large pigs. Pigs farther from the hamlets were given a per capita daily average of 0.65 kilogram of whole, uncooked sweet potato and 0.11 kilogram of scraps. The per capita daily average of sweet potato provided to all pigs was 1.3 kilograms, a figure close to that noted by Waddell. Feeding practices varied. Villagers usually gave food to pigs at the hamlets twice a day, though during the coffee season the pattern was more irregular. Pigs fed in the forest or at a remote pig house were given rations as infrequently as once a week. During 1976 and 1977, the village population had to carry approximately 200 to 400 kilograms of sweet potato each day to feed their pigs, with variations depending on the size and composition of the pig herd, seasonality in the production of gardens, the occurrence of social events, and involvement in commercial activities.

Villagers often give pigs tubers deemed too small for human

[24] During one-week periods in March, July, and November, I weighed the amount of food set aside by selected households as rations for their pigs. Four households were involved in the survey in March, six in July, and four in November. Of the 43 pigs in the three surveys, 31 were fed at the hamlets, and 12 received rations at sites farther away. I did not see the pigs consume the food but did enquire whether the weighed rations were subsequently fed to them. The data presented may possibly be higher than the amount normally fed to pigs, because people knew I would weigh their pig rations on certain days, and they may have felt "obligated" to provide the rations.

[25] Villagers feed pigs mostly raw sweet potato, but occasionally give them tubers cooked in an earth oven. They believe that pigs fed on raw sweet potato only will not grow rapidly, and that the animals need tubers cooked in an earth oven to help them grow healthy and large. Sometimes, they put ritual substances on such tubers with the aim of making pigs fatter.

consumption, though if a household's garden production is high, pigs receive larger ones as well. Women obtain many of the small tubers incidentally while harvesting larger ones for their household. At other times, they specifically collect small tubers carried by trailing vines—especially in gardens in the short, dry grasslands—just to feed the pigs.

Having enough small tubers means planting enough gardens to produce them. Villagers with several pigs must cultivate extra land to feed their domesticated animals. Some people replant in the short, dry grasslands (baeha kai'a) mainly to feed their pigs, knowing that many tubers produced will either be too small or too damaged by insect pests for humans to consume.

Pigs are not competitors for food simply because some of the harvest is fed to them; they are competitors also because they take crops when ravaging gardens. They eat sweet potato, taro kongkong, sugar cane, maize, and other foods. Thus, another essential labor expenditure in relation to pig husbandry is the need to fence gardens to keep pigs from entering and destroying them. Some villagers believe that they can escape the need to fence by cultivating very far from both the hamlets and pig houses, though pigs occasionally travel to these remote areas and devastate such unenclosed sites as well. Brookfield (1966: 55) suggested that one of the major reasons why the Simbu people to the west traditionally shifted garden sites was not so much because the soil fertility was exhausted, but because rotten fences enabled pigs to break into gardens.

Various methods besides maintaining strong fences are used in Kapanara to reduce the level of garden damage. Some villagers build pig traps or put potentially lethal, sharpened bamboo stakes where they expect that a pig will try to jump over a ditch to enter their garden. Pig owners, fearful of being asked for compensation for garden destruction and of having their pigs killed, can also take action. One of the most effective means is to keep the pigs well fed. Villagers also tie up or confine pigs at a pig house or in an enclosure for a few days or a week if the pigs have spoiled gardens. In more extreme cases, owners crop the offending pig's hooves or blind it, thereby hindering its movement.

As Table 5.6 indicates, a large number of pigs were killed because they invaded gardens. Even if his pig is only wounded by a gardener, the owner will often try to kill it; otherwise the animal might lose much weight while recovering and not be as suitable for a prestation. According to the law established by the *Eria*

Komuniti, a gardener must notify the pig owner twice about damage to his garden before killing the animal to allow the pig owner to take appropriate measures (such as confining it to a pig house) to prevent it from making further raids. However, some, enraged by the destruction they see, disregard the law and kill a pig the first time it breaks into their garden. If the garden owner does not kill the pig, he can immediately seek compensation, whether or not his garden is fenced. Sometimes the pig owner is cooperative and either pays cash, kills the pig and gives a portion of it to the gardener, or helps repair the fence and any damage to the garden. In other cases when the pig owner is uncooperative, the aggrieved gardener may take the dispute to the village court magistrate, though satisfactory results are not always immediately forthcoming.

The intensity of pressure on the agricultural system depends in part on pig herd demography. The faster the rate of natural increase, the greater the pressure. The reproductive rate of native Highlands pigs is inferior to that of Euro-American breeds, which are ready for breeding at seven to eight months, have litters of 9 to 11 piglets, and can bear two litters a year (Hide 1974: 24-25). Of the 67 sows that had litters in Kapanara in the 17-month period from July 1976 to November 1977, only 12 gave birth twice, and the average litter size was 5.91. The litter size is larger than those of 4.8, 3.6, and 3.8 reported by Malynicz (1976: 3) for other areas in the Highlands, but is not comparable to the performance of Euro-American breeds. Other demographic data are summarized in Table 5.7. The mortality rate of piglets is high. During the 17-month observation period, 486 piglets were born, but 126 pigs died, ran away, or were stolen.[26] Another 110 piglets were sold to people in other villages. Thus, low reproductive rates coupled with high mortality and numerous extra-village sales combine to limit the pressure of the pig herd on the subsistence system.[27]

The rate of increase in the pig herd is also determined by the

[26] Most of the 126 died. The majority were only a few months old.

[27] Such short-term data do not permit an assessment of whether Highlands pig herds have an inherent potential to rapidly increase and overrun resources (cf. Rappaport 1968; Brookfield 1973). Certainly the 39 percent growth in herd size from March to November, which depended almost exclusively on natural increase, is impressive. However, I do not know whether this rapid growth was atypical—resulting from seasonality in breeding or from an abnormally large number of females entering the reproductive phase—or whether it represented the normal growth pattern.

TABLE 5.7

Pig Herd Census

	July 1976			March 1977			November 1977		
	M	F	Total	M	F	Total	M	F	Total
ADDITIONS TO THE HERD									
Births				59	71	130	162	194	356
Agisted from outside the village				1	2	3	1	4	5
Purchased from outside the village				0	3	3	0	5	5
Total Increase				60	76	136	163	203	366
REDUCTION OF THE HERD									
Killings				90	99	189	87	61	148
Deaths or stolen				40	35	75	27	24	51
Agisted to outside the village				3	1	4	1	2	3
Sales				2	28	30	11	69	80
Total Decrease				135	163	298	126	156	282
Total Present	144	232	376	69	145	214	106	192	298

frequency of pig killings. Some Highlands groups such as the Simbu (Hide 1974) and the Maring (Rappaport 1968) deliberately restrict the number of pigs killed so that many large animals are available for slaughter when periodic, large-scale ceremonies are held. With such a strategy, the number of pigs is at a maximum just before the ceremony and at a minimum just after it. The Tairora do not stage such periodic, massive pig kills and thus do not employ long-term strategies restricting the killing of pigs. At least in the short term, the Tairora pattern limits the pressure of pigs on the system by continually slaughtering small- and medium-sized pigs for recurrent exchanges.

It is possible to describe roughly the form of the herd's age-sex pyramid. The pyramid is very wide at the base and narrow at the top, with an imbalance of females over males in every age grade except the youngest one. The dominance of males in the youngest class results from more females being sold, whereas the older classes reflect the greater number of male pigs killed for social and ceremonial events. Such a wide base to the pyramid is characteristic in the Eastern Highlands because villagers there frequently kill small- and medium-sized pigs for exchanges, and they do not attempt to limit fertility periodically to assist herd fattening (Malynicz 1976: 5). In contrast, other groups such as the Maring and Simbu attempt to restrict fertility and concentrate their food resources on fattening animals to produce a herd composed of many large animals suitable for slaughter at periodic festivals; the sides of the population pyramid of such pig herds are more roughly parallel (ibid. :3).

CATTLE PROJECTS AND SUBSISTENCE PRODUCTION

During my first few days in Kapanara in May 1976, I was surprised to hear what seemed to be many angry people shouting comments to others during the night. Neighbors translated their statements. The people were complaining that other Kapanarans' pigs were relentlessly destroying their gardens and that they were consequently short of food and hungry. In response to demands for compensation by the complaining gardeners, the accused pig owners denied that their animals had caused the damage, asserting that it must have been someone else's pigs. All was not well in Kapanara. Certainly, it did not appear to be the subsistence affluent economy about which I had read.

Villagers recalled that since approximately the time Papua New Guinea achieved independence (September 1975), pigs began destroying gardens at an alarming rate. Certainly the depredations of pigs are as old as pig husbandry itself, but people declared that the intensity of damage since independence was unprecedented. Villagers suggested several reasons for the level of garden destruction. A few said that people were feeding their pigs less sweet potato than usual because they were short of food, thus making the pigs hungrier. Most indicated that sorcery arising from a marital dispute was the main cause. A woman had left her husband, and when the husband demanded his marriage payment back from the agnates of his estranged wife, the agnates refused. Incensed, the husband went to see his kinsmen in the Auyana area to the southwest to enlist their support. His Auyana relatives came back to Kapanara to try and convince the woman's agnates to return the bridewealth, but when the agnates again refused, the Auyana men, who are famed for their knowledge of sorcery, recited a spell commanding the Kapanaran pigs to destroy all the village gardens. To the people of Kapanara, the incantations of the Auyana certainly appeared effective.

Complex problems do not have simple causes, and no single factor will explain the level of garden damage. Nevertheless, the building of the last four cattle projects was the major contributing factor. One hundred and thirty-four adults were involved in the endeavor, which lasted from February to April 1975. People recalled that the pace of work was not leisurely. Members of each enterprise competed against those of the other projects to complete the task. The group finishing first demonstrated its competence, skill, and strength in contrast to the other three groups. Such intergroup competition is a traditional aspect of village life and makes laborious tasks more exciting.

The tasks involved in starting a cattle project demand considerable time: felling trees, splitting fence posts, carrying the heavy posts distances sometimes greater than 2.5 kilometers, clearing a path through the grass for the fence, digging holes and setting in the posts, stapling and straining barbed wire onto the fence posts, and finally constructing a stockyard. The total work effort was extraordinary. The villagers spent 3,400 hours carrying 66,500 kilograms of fence posts, cleared 3.54 hectares of grass, erected 4,152 fence posts, and stapled 48.8 kilometers of barbed wire. The 134 adults who built the four projects worked an estimated 17,000 hours in the three-month period, an average of 42 hours per month per adult.[28]   These data indicate only a part of the labor diverted from alternative activities. People do not work continually at

[28] The estimate of the time spent in constructing the projects is based on a variety of sources. I observed and timed the labor inputs into building a subdivision fence and an extension to a boundary fence. I also recorded for all projects the number of fence posts and measured the amount of barbed wire and the length of the fences. Travelling time from the hamlets to the various tasks is based on a walking speed of 55 meters per minute. I had to rely on a rough approximation of the labor involved in building the stockyards, using estimates provided by agricultural extension officers and villagers.

Fleckenstein (1975: 132) reported that a group of Bena Bena villagers spent 1,912 man-days to construct a 7,000-meter-long, wooden post, cattle boundary fence. He does not define "man-day," but to compare his data with the results from Kapanara, I assume conservatively that a "man-day" equals five hours. Using this estimate, the Bena Bena's productivity ratio is 1.37 hours of labor per meter of fence. This fence had seven strands of barbed wire, whereas those in Kapanara have only three or four strands. Adding the extra labor cost of stapling on another three or four strands to the time that the villagers spent in building their four projects, the Kapanarans' average labor productivity ratio is a much higher 0.99 hour per meter of fence, with a range from 0.85 to 1.09 hours per meter. It is not clear whether Fleckenstein's estimate was based on informants' statements or on observation and timing of the different tasks involved. Nevertheless, the comparative data indicate that my estimate for the labor expended in Kapanara is not excessive and may even be too conservative.

arduous tasks day after day. After several days of sustained work efforts, such as those spent building the enclosures, they usually rest one or more days before beginning other laborious tasks.

Because of the demanding nature of the work, villagers reduced labor inputs into preparing and planting new gardens even though they could normally have worked in several environmental zones. The resulting food shortage caused by a decline in garden planting occurred five to six months later. This food shortage might have lasted only three to four months because people did begin to prepare gardens after finishing the projects, but the village pigs exacerbated the problem. The decrease in garden output disrupted the normal flow of rations to the pigs. Villagers generally feed pigs sweet potato tubers that are considered by many to be too small for human consumption, though much of these rations are certainly edible; for example, old men and women as well as young children sometimes eat tubers that most adults would have given to pigs. However, what is considered suitable for pigs is dependent partly on the sweet potato supply. When food is short, the pigs are first to feel the pinch as villagers consume part of the rations usually fed to their animals. Because less food was available, the pigs were fed less, and as a result the hungry swine made more determined raids into fenced gardens and went farther afield to find unfenced ones. This set in motion relationships of positive feedback or deviation amplification (Maruyama 1963): the less the pigs were fed, the more they raided gardens, which in turn meant that less food was available to feed the pigs, and so on.

Villagers' statements concerning the increased level of pig-related garden damage starting at the time of independence are understandable in light of these events. Independence was granted approximately five months after the four projects were built, the time the food shortage began, and the same time that pig rations were curtailed radically. The approximate five-month interval between project completion and food shortage reflects the lack of planting while establishing the projects; sweet potato is normally ready for harvest just over five to seven months after planting.

The food-supply problem was exacerbated by lower rainfall than normal in 1975 and 1976. Whereas the average annual rainfall at the nearby Aiyura agricultural station is 2,117 millimeters, 1,973 millimeters and 1,874 millimeters fell in 1975 and 1976 respectively. However, such deviation from the mean rainfall is not uncommon and certainly was not the major cause of the food shortage.

As an alternative explanation it may be suggested that the pig herd was too large for the people to feed. Villagers themselves reject such an explanation, declaring that their herds were larger in the past but did much less damage. In July 1976, the pig per capita ratio in Kapanara was only 0.85:1. In comparison with the pig per capita data for villages or groups in other Highlands societies, Kapanarans were certainly not caring for an abnormally large herd (see Table 5.8).

A compass and tape survey of all the gardens of the sample of households between August and October 1976 confirmed extensive damage by pigs; the survey included producing gardens as well as gardens that would have been producing during the survey period had pigs not destroyed them. Of the total cultivated area of 4.04 hectares, 23 percent was totally ruined by the pigs, an extraordinary and unprecedented amount. This survey does not reveal the full extent of the damage. Pigs often made forays into gardens, destroying a few sweet potato mounds or rooting where the trailing vines carried tubers, but they were routed by angry gardeners before spoiling much of the crop; such damage could not be measured. Complaints about the depredations of pigs were widespread. In a village-wide survey, I asked household members about the number of their gardens that were either fully or partly ruined by pigs during the period from July 1976 to February 1977.[29] At this time, most households had between four and eight dif-

TABLE 5.8

Pig per Capita Ratios of Highland Societies

Raiapu Enga[a]	2.31:1	
Simbu[b]	1.5:1	(estimated maximum adult pigs)
Mae Enga[c]	1:1	(estimate)
Kapanara	0.85:1	
Tsembaga[d]	0.83:1	
Bomagai-Angoiang[e]	0.51:1	

Sources: [a] Waddell 1972:61-62.
        [b] Brookfield and Brown 1963:59.
        [c] Meggitt 1958:288.
        [d] Rappaport 1968:14,57.
        [e] Clarke 1971:19,84.

[29] Ninety-three percent of the Kapanaran households were sampled to obtain the data. The survey, of course, does not reveal the number of gardens invaded from the beginning of the hunger period, September 1975, to June 1976.

ferent producing gardens. Eighty-one percent of all households had at least one garden damaged by pigs (see Table 5.9).

Pigs spoiled both fenced gardens and unfenced ones far from the hamlets. The level of damage is further indicated by the fact that of the 120 pigs killed from June to October 1976, 50 percent were slaughtered because they ruined gardens. In contrast, Hide (1974: 34), using a smaller sample in a Simbu community, noted that between May 1972 and April 1973 only 4 percent of the pigs were killed because of garden invasions. The abnormally high rate in Kapanara reflected the impact of a strong commitment to commodity production.

The food shortage in Kapanara lasted from late 1975 until the end of 1976. No one starved, but many complained of hunger, and some had to rely on gifts of food from relatives in nearby villages. The problem was ameliorated partly by killing a large percentage of the pig herd, thus accounting for the decline in the herd size between July 1976 and March 1977 (see Table 5.7). The institution of the village court also affected the herd reduction because people feared recurrent fines by the village magistrate each time their pigs ravaged a garden, and thus several decided to kill some of their own pigs to avoid compensation claims.

The hypothesis of Vayda, Leeds, and Smith (1961) must be examined in light of these events; they suggested that pig husbandry in the Highlands is adaptive because it is a means of storing vegetable surpluses that can be used in times of food shortage. Certainly, pigs are food reserves on the hoof. However, a critical element in their usefulness in this role is the timing of people's reaction to reductions in the food supply. Human populations do not have behavioral mechanisms that act as perfect homeostatic

TABLE 5.9

Number of Gardens Damaged by Pigs

Number of Gardens Damaged	Number of Households
0	19
1	41
2	25
3	9
4	4
5	2

regulators to change the state of some components in the ecological system in immediate response to only slight perturbations in others. Time lags always occur, and furthermore, when commodity production is introduced, many traditional patterns of coping with stress are disrupted. In this case, Kapanaran villagers were reluctant to kill their animals in times of shortage because pigs were highly valued both as a source of income and as an integral part of the traditional exchange system. Although pigs technically were food reserves on the hoof when the food shortage began, villagers did not kill them fast enough to control the problem. Consequently, pigs destroyed more food than they themselves provided on the hoof.

Although I have placed major emphasis on the detrimental effect of allocating so much labor to building the projects in a relatively short time, other changes in productive activities also contributed to the food-supply problem. When the four new cattle projects were stocked in November 1975, many households had producing gardens within the area enclosed by the new boundary fences. Twelve households abandoned such gardens within the next few months either because cattle had destroyed their gardens or because they did not replant them, fearing that the cattle would eventually ruin their crops. They preferred to plant new gardens elsewhere rather than recultivate in the enclosed areas. Tilling and replanting an established garden are much less time-consuming than preparing an entirely new one, and thus these households experienced further delays in food production. Another contributing problem was the labor input into harvesting coffee in the village and picking coffee in a nearby plantation; both activities were stimulated by abnormally high coffee prices in 1976. Such inputs were made at the expense of garden work and producing enough food for the pigs (see Chapter Six).

The labor input into establishing the projects created problems temporary in nature. Government policy in the 1970s, by influencing the size and location of the projects, has had a more lasting effect on subsistence production. DPI required that all projects built since the late 1960s enclose all land used for cattle grazing within a barbed wire boundary fence. The PNGDB preferred lending to projects considered large enough to be economically viable—those with the capacity to support at least 10 to 15 breeders and that were 60 hectares or larger, though smaller projects also received funding. Agricultural extension officers also promoted large projects because they felt that a multitude of smaller ones would be too difficult to service. They thus informed villagers of the importance of fencing relatively large areas. Extension officers also persuaded people in Kapanara to build the enclosures near the road to make delivering cattle or providing veterinary services easier for the officers. As the hamlets in Kapanara were near the road, they were also near the cattle projects (see Map 1.5).

Government policies were not the only factors influencing the size and location of the projects. The government's emphasis on relatively large, fenced projects was consistent with the desires of the villagers involved. The people realized that the larger the area enclosed, the bigger the loan they could obtain and hence the more cattle that would be delivered. That would, in turn, earn them more prestige and income.

As a result of government policies and the desires of those starting the enterprises, the seven village cattle projects enclosed a considerable amount of grassland near the hamlets—5.49 square kilometers or 28 percent of the entire grassland area in the village territory. The ground taken up by the projects is not only steep, well-drained *Themeda* grassland of limited agricultural value. Villagers also fenced prime agricultural land in the valley bottom for several reasons. Cattle need water to drink, and water is most readily available in streams in low-lying areas. Furthermore, villagers observe the feeding behavior of cattle and believe that cattle prefer to eat *Ischaemum polystachyum*,[30] the dominant fallow vegetation in parts of the valley bottom. Also, digging fence post holes is much easier in moist-to-swampy soils than in the more compacted soils of the well-drained hillsides, and fence posts last longer in relatively wet soils because they are not constantly exposed to alternative periods of wetting and drying, which rot the hillside posts more rapidly. Besides fertile, valley bottom soils, good land on the hillsides that is suitable for growing sweet potato, yam, and winged bean (*baeha kai'a*) as well as taro (*ha'i kai'a*) has also been fenced.

Previous gardening patterns reveal the agricultural utility of the enclosed land. Eighty-three percent of the Kapanaran households cultivated at least one site within one or more projects before the enterprises were established (see Table 5.10). However, the data in Table 5.10 do not reveal the true extent of the value of the enclosed land because many individuals previously cultivated more than one site within a particular project. In addition, a number

---

[30] *Ischaemum* is the dominant fallow grass where villagers engage in *hora kai'a*.

TABLE 5.10

Previous Use of Land for Gardens in Area Enclosed
by the Cattle Project Fences

Number of Projects Enclosing at Least One Previously Cultivated Site	Number of Households
0	17
1	22
2	16
3	16
4	15
5	5
6	3
7	4

of those listed as never cultivating such land were married after many of the projects were constructed; unmarried individuals prepare few gardens.

Although the land is valuable agriculturally, few people cultivate inside the enclosures. Cattle bosses restrict the use of land within their project, and villagers themselves are hesitant to cultivate inside the enclosed areas. To prevent cattle from ruining their gardens, people must build much stronger fences than those designed to keep out pigs. Cattle have destroyed many of the gardens inside the boundary fences. Consequently, few individuals prepare new food gardens within the confines of the projects.

In reaction to the large area enclosed near their residences, villagers prepared more gardens at a greater distance from their hamlets in the mid-1970s than previously. In a review of distances to garden sites in Melanesia, Brookfield with Hart (1971: 225) stated:

In Melanesia, in areas where only foot transport is available, we can empirically suggest that, as a norm, about half the garden land of a community, or of individual farmers in dispersed settlements, will be found within 1.0 km of the residence, while we will often find at least 75 per cent within 2.0 km.

Brookfield (1973: 149-150) later observed:

A distance range of 1,000-1,500m crops up rather frequently in Chimbu. Within it lie the range from residence within which lie 75 per cent of gardens. ... In Chimbu, as in many other societies, there is a strong suggestion of a 'tolerable distance' limited somewhere within this range, and beyond which the separation of everyday activities becomes onerous and unusual.

In Kapanara, only 38 percent of the gardens were within 1.0 kilometer of the residences and 59 percent within 2.0 kilometers (see Table 5.11). The difference between the norms suggested by Brookfield with Hart (1971) and Brookfield (1973) and the distribution of gardens in Kapanara is due largely to the presence of the cattle projects. In addition, some villagers maintained that they wanted to cultivate far from the hamlets to escape the pigs, though the pig-depredation problem was also related in part to the cattle projects.

The increase in distance to the gardens had several implications. First, the energy expended on travelling to gardens and carrying home the food harvested, a burden falling mainly on the women, increased considerably. The elderly especially complained of the problem of having to travel farther to obtain food. Second, as a rule the quality of garden care as evidenced by the thoroughness of weeding and fence maintenance generally declines with increasing distance from the hamlets, and thus garden care suffered. Third, more gardens were now located in the zone of maximum danger from sorcery and the related *sangguma* attacks by mem-

TABLE 5.11

Percentage of Total Area of Subsistence Gardens at Varying Distances from the House Sites, Sample of Households, 1976

Distance (m)	Percentage of Total Area Cultivated
0-100	20
101-500	12
501-1000	6
1001-1500	18
1501-2000	3
2001-2500	23
2501-3000	9
3000+	9

bers of other villages. Sorcery danger has spatial dimensions. It is felt that the possibility of a sorcery attack is greatest in areas farthest from the hamlets. Sorcerers prefer to attack far from the hamlets to lessen the chance of their being discovered; the closer a place is to the hamlets, the more likely Kapanarans will discover the hiding sorcerer and either kill or rout him. Members from other villages practicing sorcery and *sangguma* are believed to hide and wait to attack a lone individual. Prime victims of such attacks—women, children, and the elderly—often refuse to go alone to distant gardens because of fear of a sorcery or *sangguma* attack. After the death of an individual from a nearby village results in accusations that Kapanarans committed lethal sorcery, men also do not travel alone to faraway gardens for fear of retaliation. Occasionally, people who wanted to go to their distant gardens did not do so because they could not find anyone else to go with them.

With so much valuable garden land taken up by the enterprises and the resulting spatial consequences, it is understandable that when a group attempted to begin yet another cattle project in early 1976, the rest of the villagers objected vociferously. The people demanded that the remaining unenclosed land be left for gardening activity. However, there was not a shortage of land *per se*. Even with the presence of the cattle projects, Kapanarans considered that ample land for gardening remained in the village territory, a viewpoint with which I agree. The problem was the distance they had to travel to their gardens.

Cattle also became a problem as direct competitors for food. Villagers did not always properly maintain the barbed wire boundary fences, and as a result cattle occasionally escaped from the confines of the projects, wandered through the village, and destroyed food gardens. Cattle ate the leaves of various crops, knocked them over and trampled them, and compacted the garden soil. This problem was not confined only to Kapanara but was widespread throughout the Highlands. In fact, the problem was so extensive that agricultural extension officers in several provinces published lists of standardized compensation payments for various crops destroyed by cattle. Using the compensation list established by Kainantu extension officers, an anthropologist working near Kainantu reported that on one occasion cattle from a large project damaged over K1,600 worth of crops (G. Westermark, personal communication 1977). Although such a level of damage represents an extreme, it does indicate the potential impact of cattle.

CHAPTER SIX

Time and Money in
the Mid-1970s

SMALLHOLDER CATTLE PROJECTS were not the only aspect of the village commercial economy that affected local food production. The intense Kapanaran commitment to other commercial endeavors in the mid-1970s—particularly coffee production—also had profound implications for the viability of subsistence production. The following analysis of these other commercial activities highlights two additional factors affecting the impact of commodity production. One is seasonality in the allocation of time to subsistence and cash-earning endeavors, and the other is the nature of rural expenditure patterns. Seasonal conflicts between the two realms clearly existed in the mid-1970s, and Kapanarans' devotion to spending large amounts of money on beer drinking and gambling had severe negative consequences for food gardens and pigs. Thus, to provide a more thorough assessment of the conflicts between subsistence and commodity production, we must also consider other income-producing activities as well as patterns of cash expenditure.

COFFEE PRODUCTION

Kapanarans' commitment to coffee production has varied considerably through time, with periods of enthusiasm in coffee planting and harvesting depending largely on price fluctuations. When coffee prices are low, labor inputs into coffee production are minimal and a portion is left unharvested—a pattern widespread in the Highlands. In contrast, when prices are very high, villagers pick all berries on the trees and even collect beans that have fallen to the ground. Starting in late 1975, the price rose dramatically in response to frosts in Brazil, civil war in Angola, earthquakes in Guatemala, and coffee-rust disease in Nicaragua, all of which reduced the supply of coffee on international markets (Townsend 1977: 419). Thus, my fieldwork in 1976 and 1977 coincided with a period of abnormally high coffee prices (see Figure 6.1) and

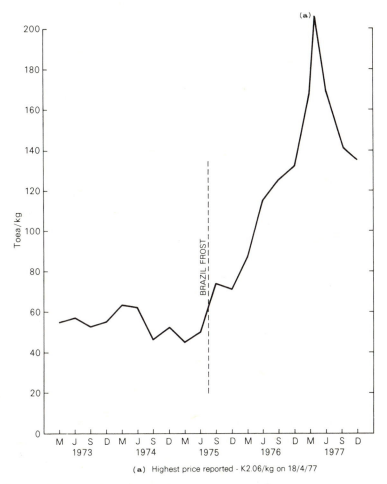

Figure 6.1. Factory Door Price for Dry, Sound Quality, Arabica Parchment Coffee, 1973-1977. *Source*: Papua New Guinea Coffee Industry Board, Goroka.

intensive inputs into coffee production. It provided a prime opportunity to observe first-hand the impact of an enthusiastic peasant commitment to commodity production—the type of commitment that planners and government bureaucrats are eternally seeking.

Coffee production is an individual or household enterprise and thus contrasts sharply with the initial communal organization of most cattle projects. Every married household owns some coffee, and hence coffee income is distributed widely in the community. Individuals obtain plots mostly by planting and to a lesser extent by gift transfer or inheritance. Although both men and women can own separate coffee holdings, husband and wife usually harvest together from the same plots and pool their income. However, some women prefer keeping proceeds from the sale of coffee that they picked to prevent their husbands from spending too much of the household's coffee income on beer and gambling. In polygynous households, each wife works her own plots.

Coffee production also contrasts with cattle raising because it is not dependent on external credit sources. Villagers do not employ pesticides or fertilizers, and thus the financial outlays involved are minimal.

Most of this perennial tree crop in Kapanara is near the hamlets, reflecting both the intensive labor inputs into coffee production during the flush and the need to guard against theft. Eighty-eight percent of the coffee was within 1,000 meters of the house sites (see Table 6.1).[1] Data from other Highlands societies reveal a similar concentration near residences (see Brookfield 1968a: 103; Hide 1975: 13).

Because Kapanarans plant coffee in gardens ready to be fallowed and next to house sites, the distribution of a household's plots is fragmented. Married households in the sample owned between

TABLE 6.1

Percentage of Total Area of Coffee at Varying Distances from the House Sites, Sample of Households, 1977

Distance (m)	Percentage
0-100	18
101-500	7
501-1000	63
1001-1500	10
1501-2000	0
2001-2500	0
2501-3000	2

[1] In the sample of households in Kapanara, most coffee was located between 501 meters and 1,000 meters from the house sites and not, as one might expect, within 500 meters. This pattern reflects past residence changes after coffee was planted. Of the coffee in the 1001-1500 meter range, the farthest plot was 1,218 meters from a house site. The one plot located over 2,501 meters from a residence was obtained as a gift transfer and not planted by the present owner.

two and eight coffee gardens (with a mean of five) in different areas of the groves. Individual plots ranged in size from a few trees up to 0.2 hectare.

Married households owned between 0.09 and 0.46 hectare, with polygynous units having the largest holdings.[2] The elderly had the least coffee, because they had already transferred most of their plots to their children; the three widowed individuals in the sample owned between 0.02 and 0.04 hectare. For Kapanara as a whole, the coffee groves only occupied approximately 19.7 hectares in 1977, or 0.045 hectare per capita.[3]

A few individuals in the village possessed more coffee than those in the sample, but such holdings were not substantially larger than the figures presented here. Although differences in the amount of coffee owned in the village clearly exist,[4] they are not nearly as marked as those reported elsewhere in the Highlands (see Finney 1973; Gerritsen 1979; Good and Donaldson 1980) and are certainly not as crucial in facilitating economic differentiation as the cattle projects. Unlike cattle raising, coffee production is not restricted by unequal access to land, because land is still available for those wishing to expand their holdings. All married households in the sample, for example, were establishing new coffee plots in 1977, with the amount planted determined by initiative and available household labor.

Just as the size of holdings differ, there also is variation in husbandry practices. Most growers provide a low level of care that constrains yields, a condition common in the Highlands (see Howlett et al. 1976: 220). Lack of technical knowledge, periodic disillusionment with coffee prices, and the demands of other activ-

[2] Data on coffee gardens for the sample are based on compass and tape surveys. The figures include both bearing and new trees. The largest holding of bearing trees was 0.3 hectare.

[3] This estimate is based on several sources. For the sample, the area of coffee per capita was less, 0.040 hectare. For comparative purposes, I estimated the extent of the village coffee gardens using a 7.5x enlargement of a Royal Australian Air Force 1974 aerial photograph at a scale of 1:100,000. The data obtained from the 1974 aerial photograph had to be supplemented because additional coffee was planted between 1974 and 1977. As the area of coffee cultivated by the sample increased by 40 percent since 1974, a similar increase was estimated for the holdings of the village population. Using this measure, the village groves are estimated to have occupied 19.7 hectares of land in 1977, or 0.045 hectare per capita, a figure only slightly higher than that for the sample.

[4] Although most cattle bosses owned more coffee than the village average, a few did not. Some others who were not bosses also had relatively large holdings.

ities account for poor husbandry practices, which are evident in all phases of coffee growing. The quality of planting material is not uniform. Occasionally, villagers obtain high-quality seedlings from agricultural extension officers, but more frequently, they transplant seedlings that have sprouted spontaneously in the groves; hence planting material is not necessarily selected from the highest-yielding trees. In the coffee groves, villagers often grow *Casuarina* trees, which reduce weed competition, provide shade, and thus facilitate nutrient uptake. However, not all groves are shaded, and unshaded coffee produces poorly, particularly on dry, well-drained hillsides. Because coffee is a surface-feeding plant, weeding is essential to ensure maximum production (Barrie 1956: 23), but attention to weeding is minimal. Most people wait until February or March to begin removing weeds that have grown during the rainy season. Both men and women weed their plots with the aid of a bushknife a few times during the coffee-harvesting season, but after the flush, attention to weeding is again negligible. Pruning to remove excess and unproductive branches and to produce multiple-stem trees increases production. Many villagers do make at least minimal efforts at pruning, and many multiple-stem trees can be found in the groves; nevertheless, further attention to pruning, especially at the end of the flush, would increase yields.

Although coffee bears berries throughout the year, most of the yield occurs in a flush from May to August, and thus villagers make most of their labor inputs into coffee production during this period. The various stages in coffee production are harvesting, pulping, fermenting, washing, and drying. The sexual division of labor in these tasks is not marked, except in the case of washing coffee, which women do much more frequently.

One of the most familiar sounds during the coffee season is the "ping" of hand-picked berries being dropped into a metal bucket. During harvesting, household members—men, women, and children—frequently work together. They largely harvest their own crop, though occasionally they may receive help; for example, the elderly sometimes assist their children in picking coffee to reciprocate for food that they receive. Much more rarely, those with large holdings either employ children at the rate of 20 to 50 toea per day, or else they invite other adults to help and give them a part of the harvest. The work pace is never intensive. People owning adjoining plots often take short rests together, sitting around a small fire, smoking cigarettes, and talking. After the work of harvesting is finished, they gather their picked berries, pour them in a sack, and carry them home.

Within one day after harvesting, villagers remove the red coffee skin and fleshy pulp covering the beans. Households that have contributed to the purchase of one of the seven hand-operated pulping machines in Kapanara use it for the task; they pour coffee and water into the top of the machine and turn a handle driving a sharpened drum that removes the covering. Because mechanical coffee pulpers are expensive (K140-K170 in 1977), individuals—usually between five and fifteen men—pool money to purchase one, and as a rule only the contributors use it free of charge. The one or two largest contributors take charge of the machine[5] and erect it near their house. Because most of the hamlets and coffee are located in the same area, the distance that other contributors must travel to a machine is never very great.

Villagers who do not contribute money but who want occasionally to use a machine normally pay a small fee—50 toea, K1.00, or more—to the person in charge, though sometimes close kinsmen are permitted to use it without cost. Alternatively, those who did not help purchase a machine remove the covering by the more time-consuming methods of chewing or using a hammer or stone to lightly pound berries placed under a cloth. Thirty-one percent of the households had not previously been contributors and hence often relied on these methods.

After pulping, villagers put the sticky beans in a small, wooden box approximately 70 centimeters long and 40 centimeters wide, and the fermentation process begins. Enzymes start to break down the sugary mucilage coating the bean. Ideally, fermentation should continue for two days, but some villagers allow the coffee to ferment for one day only, which is insufficient time to fully break down the mucilage covering the bean and makes subsequent drying more difficult (J. Yogiyo, personal communication 1977).

Then women pour water into the box to wash the coffee and remove the pasty coating. When finished, they place it on a blanket, *Pandanus* leaf mat, or raised bamboo platform next to their house to dry in the sun. The drying phase involves the greatest risk, because a few village youths occasionally attempt to steal unguarded beans. If one of the household members is not near the residence, neighbors or elder relatives sometimes look after

[5] Cattle bosses contributed the largest or second largest amount in the purchase of six of the seven pulpers, indicating their relatively high income level.

the drying beans, but on other occasions the coffee is left unattended. During the drying period, people bring their coffee in each night and set it out again the next day if rain does not threaten. Depending on the sun's intensity, villagers usually leave the beans to dry for three or four days or longer. Much coffee sold is dried only partially and thus commands a lower price per kilogram than fully dried coffee. Complete drying may take up to ten days in the sun (PNG, Coffee Marketing Board 1972: 7).

People sell most of their coffee at the village road to itinerant, national coffee buyers.[6] The buyers are independent agents who, in turn, sell to coffee-processing factories in Kainantu. Villagers can receive approximately 20 to 40 percent higher prices for their coffee in Kainantu, but many prefer the safety and convenience of selling in Kapanara. In addition, many PMV operators charge passengers extra money to bring coffee to Kainantu, thus reducing the potential benefit of selling there. Consequently, only 22 percent of the villagers' coffee income in 1977 was obtained by selling in town.

When a coffee buyer arrives in his truck, villagers with sacks of coffee to sell, as well as additional spectators, gather on the road. One of the buyer's helpers places a sack on a hand-held scale, calls out the weight, and then the buyer sitting in the cab of the truck makes the villager a cash offer for his coffee. People often compare prices offered by different buyers, and they will not sell to one buyer if they believe that they can obtain a better deal from another. Prices offered by different buyers may vary up to 20 percent or more on a single day. Relations with coffee buyers are often strained because many villagers believe that the buyers cheat them and are largely responsible for price fluctuations and low prices (see also Strathern 1982b). Certainly, part of the problem lies in international fluctuations in coffee prices, which are beyond the buyers' control. However, a few coffee buyers do occasionally act dishonestly, for example, in understating purposely the weight of a coffee sack or its appropriate price.

For the thirteen sample households, the gross coffee income in 1977 was approximately K2,450[7] (K188 per household and K45

---

[6] Since 1974 it is illegal for Europeans to purchase coffee directly in the villages.

[7] This figure is based largely on fortnightly records that I kept from March to December 1977. In addition, in March I asked them how much coffee they had sold since the beginning of the year; the amount was small, because little coffee ripens between January and March. This figure was incorrectly reported as K2,600 in Grossman (1979).

per capita), making this tree crop their major income earner. As sample members owned 1.59 hectares of bearing trees, the return per hectare was K1,540. Coffee income, as it generally follows the flush, is highly seasonal, though villagers sometimes store dried coffee in their house to obtain cash later for important social occasions such as Christmas (see Figure 6.2).

WAGE LABOR

Kapanarans first earned cash wages in the 1950s when working as carriers for administration patrol officers. Recruited under the now defunct, administration-sponsored Highland Labour Scheme, men also worked as contract laborers for periods of two years or more on lowland plantations. In the mid-1970s, picking coffee at Highlands plantations and working in the village as road laborers for the Public Works Department were their two main forms of wage labor.

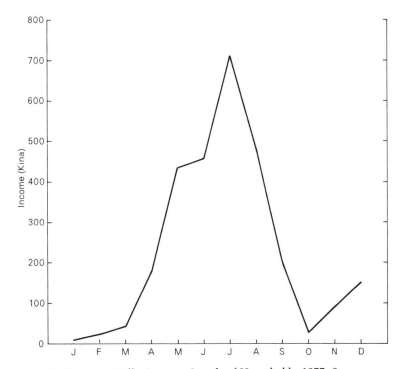

Figure 6.2. Kapanara Coffee Income, Sample of Households, 1977. *Source*: Grossman 1982.

In 1977, the government employed on average six Kapanaran men to help maintain the road passing through the village. Some men kept their job for the entire year, whereas some worked for only two or three months before being replaced by others. Working Monday to Friday, from about 8:00 a.m. to 4:00 p.m., the road crew weeded the roadside and maintained the unpaved road surface and drainage ditches. The government paid each worker K22.70 per fortnight, and as an average of six men were employed throughout the year, the village earned K3,540 in 1977. One member of the household sample worked as a road laborer.

In terms of the number of people employed, harvesting coffee at the nearby plantation at Noraikora was the most popular form of wage labor in 1976 and 1977. Villagers were paid according to how much they picked, and as the amount they received for each kilogram harvested depended in part on the coffee price, wages at the plantation in 1976 and 1977 were very high compared to previous years. Thus, more Kapanarans than usual sought work there. Husband and wife sometimes worked together, and several children from different households commonly combined their harvests and shared the income. When employment was available, villagers could work as many or as few days as they wished, with most choosing not to work a full week to leave time for various tasks and activities, such as picking and processing their own coffee. Because the plantation is nearby, people were able to return home to Kapanara each day after work.

In 1977, Kapanarans' total wages at the plantation were K1,479 (see Table 6.2). Married adults were the main income earners, with men receiving the larger proportion of money. The level of village employment in 1977 was significantly lower than that in 1976, when Kapanarans earned K3,292; because the plantation owner had a disagreement with some Kapanarans in 1977, he restricted employment for most villagers to a few fortnights in August and September. However, many were eager to work at the plantation because of the relatively high wages, and had there been no restrictions, villagers would have earned considerably more money there in 1977.

To supplement their wage labor in 1977, some adult men, most of whom were unmarried and had little coffee at home, worked at other more distant plantations in the Eastern Highlands Province. Although these men collectively earned K408 (see Table 6.2), they probably brought back no more than K100 to the village

TABLE 6.2

Wage Labor at Coffee Plantations

(a) Noraikora Plantation

	1976												1977											
	Married Adults				Unmarried Adults 15 Yrs or Older				Children below 15 Years				Married Adults				Unmarried Adults 15 Yrs or Older				Children below 15 Years			
	M		F		M		F		M		F		M		F		M		F		M		F	
	I	FW	I	FW	I	FW	I	FW	I	FW	I	FW	I	FW	I	FW	I	FW	I	FW	I	FW	I	FW
	1358	137	741	93	751	61	199	25	74	16	169	28	757	70	346	37	227	26	54	8	44	12	51	9

Total income = K3292
Total number of fortnights worked* = 360

Total income = K1479
Total number of fortnights worked = 162

(b) Other Plantations, Eastern Highlands Province, 1977

Married Males		Unmarried Adult Males, 15 Years or Older	
I	FW	I	FW
72	7	336	37

Total income = K408
Total number of fortnights worked = 44

Key: M = Male F = Female I = Income FW = Fortnights worked
* The term 'fortnights worked' does not mean how many complete fortnights individuals worked but only on how many occasions they received wages.

because they received low wages (K10 per fortnight or less) and had to purchase part of their food while away overnight.

Most married adults who worked at the nearby Noraikora plantation were younger than 35; married adults with more extensive coffee holdings, usually older people, remained at home. Several would not seek employment there fearing that the Noraikora people, Kapanara's traditional enemy in the precontact era, would try to kill them by sorcery or *sangguma* attacks while at work. During the period from 1976 to 1977, one cattle boss did engage in wage labor at the plantation but only for one week in 1977. Project leaders generally prefer to work their own coffee holdings. However, bosses were not adverse to wage labor. The one sample member who was a road laborer was a cattle boss, and another project leader joined the road crew in late 1977. Wage labor in the village is perceived as being much safer, and if employment begins after the coffee flush, it does not conflict with harvesting one's own coffee, unlike work at the plantation.

MARKETING

Since the 1960s, villagers have been selling some of the surplus from the subsistence sector at the market place in Kainantu. They market produce to supplement their income as well as offset the expense of coming to town for other reasons, such as visiting relatives, selling coffee, or purchasing beer and packaged foods. The Kainantu market is operated by the Kainantu Local Government Council, which charges vendors a small entrance fee. In this outdoor market, sellers sit mostly at tables, displaying fresh produce, betel nut and lime, and, to a lesser extent, clothing and cooked items. Customers, who are mostly nationals, include both government employees living in town and people from villages in the region.

Earnings from marketing in 1977 were highly seasonal (see Figure 6.3),[8] with people engaging in most of their marketing in the first three or four months of the year to offset low coffee income during this period. Villagers also concentrated their produce-selling activities early in the year because it is the time when nuts from the *Pandanus* tree and *Areca* sp. palm ripen. Both nuts are

[8] The figures for January and the first week of February in Figure 6.3 are estimates. By collecting information fortnightly from sample members, I obtained the data for the rest of the year.

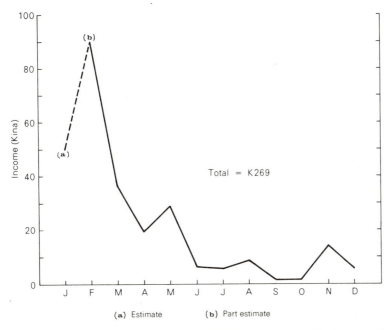

Figure 6.3. Kapanara Income from Selling Produce at Kainantu Market, Sample of Households, 1977

popular items in the town market. For the entire year, smoked tobacco leaves and peanuts were the largest revenue earners (see Table 6.3).[9]

Villagers preferred to market on Thursdays, when government employees received their salary (see Table 6.4). The produce that a household sold was marketed largely by one of its own members. Vendors, who were mostly women, often sold more than one type of food during a visit to the market. The average gross income for a household on a single day was K4.07, with a range from no sale to K13.70.

Considering the expenses of transportation and the small entrance fee, the net profit per day for a household engaged in marketing averaged K2.95, with a range from zero to K12.00. Six visits

[9] The relative proportions of all items sold as listed in Table 6.3 are representative for the entire year, except those for *Areca* palm nuts and *Pandanus* nuts because of seasonality in production. Villagers sold a considerable amount of both in January, and thus the figures are underrepresentative. Some Kapanarans sold other produce such as sugar cane, but no one in the sample did.

TABLE 6.3

Gross Revenue from Various Items Sold in Kainantu Market, Sample
of Households, February to December 1977

Item	Sales (in Kina)	Percentage of Total Sales
Smoked tobacco leaves	73.16	35
Peanuts	49.10	23
Sweet potato	18.20	9
Cabbage	17.90	8
Areca palm nuts	16.00	8
Maize	13.20	6
Winged bean	7.90	4
Pineapple	3.20	2
Taro kongkong	2.60	1
Bananas	2.30	1
Pandanus nuts	1.90	<1
Lettuce	1.70	<1
Cucumber	1.00	<1
Leaf of *Piper wichmannii* (*Lip daka* in *Tok Pisin*)	1.00	<1
Oenanthe javanica	1.00	<1
Other	1.40	<1
Total	211.56	

out of 52 resulted in transport costs and entrance fees either equal-
ing or exceeding the cash returns from sales, though such a situation is not necessarily a loss from the villagers' perspective
because marketing is often used to offset the expense of coming
to town for other reasons. Members of the sample earned a net
total of approximately K190 from marketing produce in 1977.
There were no major differences between the marketing earnings
for the households of cattle bosses and for the others.

The sale of home-grown produce to others within Kapanara was
much less prevalent than marketing in Kainantu. Most intravillage transfers of food and produce grown in Kapanara were still
made within the context of the traditional system of reciprocal
exchange. However, some villagers occasionally sold such items
as sugar cane, bananas, and pineapples to those playing cards or
at various festivities. Although the commercialization of relations

TABLE 6.4

Distribution of Days When Produce Is
Marketed in Kainantu, Sample of Households

Days Marketed	Percentage of All Days Produce Sold
Monday	17
Tuesday	10
Wednesday	17
Thursday	41
Friday	3
Saturday	10
Total	98*

* Less than 100 due to rounding.

within Kapanara is only at an incipient stage, the sale of produce to fellow villagers nevertheless indicates a trend toward increasing individualism and a significant expansion of the commercial economy in rural life (see also Nietschmann 1973, 1979).

SUMMARY OF VILLAGE INCOME

Selling coffee, cattle, and pigs; marketing produce; and working for wages were the five major sources of income in 1977. The estimate of village revenues is based partly on marketing and coffee income data extrapolated from the sample to the entire village population.[10] Data on wage labor and pig sales are based on village-wide surveys. Revenues from the sale of cattle are not included in the computations because villagers spent K2,620 repaying their PNGDB loans, whereas they earned only K1,710 selling cattle to members of other villages. The dependence of the village economy on seasonal coffee revenues is evident (see Table 6.5).

As much of the data in Table 6.5 is derived from villagers'

[10] The coffee income figure is derived by assuming a return per hectare of K1,540. As there were approximately 15.3 hectares of producing coffee trees in Kapanara, the village coffee income for 1977 is estimated to be K23,600. To obtain the result for marketing, the ratio of the number of fully active wives in the village to those in the sample (8.15:1) was multiplied by the earnings of the sample, K190, to yield an annual village marketing income of K1,530.

TABLE 6.5

Sources of Village Income, 1977

Source	Income (Kina*)
Selling coffee	23,600
Wage labor	5,100
Sale of pigs	3,100
Marketing	1,500
Total	33,300

* Rounded to nearest hundred.

statements and extrapolation from the sample, the approximate nature of the results must be stressed. Nevertheless, the data are useful as indicators of the relative contribution of different cash-earning activities to overall village income and of the importance of the commercial economy in general.

THE CONSUMPTION OF ALCOHOLIC BEVERAGES

Much of the money earned from the sale of coffee is spent on alcoholic beverages, a relatively recent phenomenon. The people of Kapanara, like other Highlanders, have no traditional alcoholic beverage, having adopted the custom of drinking beer and liquor from Europeans. Although the Australian administration first granted nationals permission to drink alcoholic beverages in 1962, this concession had negligible impact on Kapanara in the 1960s. Villagers had their first drinking experiences while working as plantation laborers on the coast and later at *singsings* in the Kainantu area. However, they did not start consuming large amounts of beer and liquor in their own village until 1972 or 1973. Since then, beer drinking has rapidly become a central feature of their social and ceremonial life.

According to the law, alcoholic beverages must be purchased in Kainantu town either at the licensed hotel or at one of several licensed outlets. Because of the extensive road network in the District, most villagers, including the Kapanarans, have direct access to town and therefore to alcoholic beverages as well.

Kapanarans have a strong preference for beer over hard liquor, and approximately 90 percent of the money they spent on alcoholic beverages in 1977 was used to purchase beer, which, at that time, was available only in cartons of 24 bottles or cans.[11] The price per carton of bottled beer varied during 1977 in response to changing government regulations (see Table 6.6). The price rise on July 15, 1977, reflected a liquor sales tax of 48 toea per carton of beer imposed by the Eastern Highlands Provincial Government. As an indication of the extent of beer drinking in the Province, the tax quickly became the major internal income earner for the Provincial Government (PNG, *Post-Courier*, September 20, 1977, p. 9).

The Kapanaran community consumed a total of approximately K10,000 worth of beer in their village in 1977, with the monthly amount spent on purchasing cartons of beer varying considerably (see Figure 6.4 and Appendix C.1) (Grossman 1982). People also paid approximately K1,000 for hard liquor brought to Kapanara and another K500 for beer consumed at the licensed hotel in Kainantu. Thus, at the very least, villagers spent K11,500 on alcoholic beverages in 1977, or 35 percent of the entire village income. Similarly, Townsend (1977) estimated that in 1976, 29 percent of Highlands coffee-related income was spent on beer. The large amount of money allocated to purchase beer is typical of many Tairora villages but not of other groups in the Kainantu District.

Coffee sales and beer consumption in 1977 followed the same seasonal trend except in October (see Figure 6.2 and Figure 6.4). More beer is purchased during the coffee flush from May to August

TABLE 6.6

Changes in the Price of a Carton of Bottled Beer, Kainantu, 1977

Period	Price (Kina)
January to June 1	8.55
June 2 to July 14	8.05
July 15 to August 3	8.50
August 4 to December 1	9.05
December 2 to December 31	9.10

Source: Grossman 1982.

[11] A carton of beer contains twenty-four 0.33 liter bottles. People preferred bottled beer over cans because canned beer was between 75 toea and K1.10 more expensive per carton.

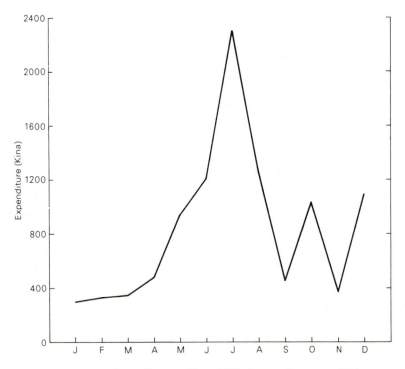

Figure 6.4. Kapanara Expenditure on Beer, 1977. *Source:* Grossman 1982.

when more money is available than at other times of the year. In addition, village income is augmented during the flush because most payments from people in other villages for the purchase of Kapanaran piglets are made during this period. The increase in beer purchases in October may reflect the sharp rise in wage labor in September 1977 at the nearby coffee plantation. The increase in December definitely is related to the Christmas season. In December, villagers stockpile beer for various parties planned for the Christmas holiday, a time when people rest from arduous activity. The general relationship between the sale of coffee and the purchase of beer also is revealed in the records of the Kainantu Local Government Council, which both sells beer to and purchases coffee from people in the Kainantu region (see Figure 6.5).

Consumption of alcoholic beverages is now an integral part of village social life. Both men and women drink, though men do so more heavily. Men begin to participate fully in drinking in their late teens. Usually only women who are or were married indulge. Most adults, even the elderly, drink at least occasionally.

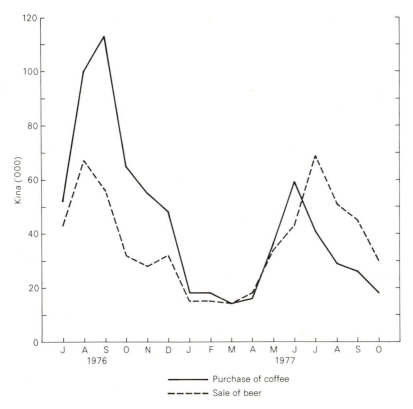

Figure 6.5. Beer Sales and Coffee Purchases: Kainantu Local Government Council, 1976-1977. *Source:* Kainantu Local Government Council.

Even children as young as four sometimes consume beer, a phenomenon that occurs elsewhere in the Eastern Highlands (see Sexton 1982). Almost all drinking occurs in or near the hamlets.

Villagers give cartons of beer and bottles of liquor—along with pork, beef, vegetables, and store-bought foods—in formal prestations such as affinal and *kandere* exchanges, in mourning ceremonies, and to reciprocate various services such as assistance in preparing gardens, building houses, and fighting. If beer is not provided on these occasions, the event is felt to be less enjoyable and inferior. The most cartons exchanged in a single transaction during 1976 and 1977 was 24, when two brothers each gave the other 12 cartons of beer to signify the settling of a dispute. However, in the mid-1970s, more than one event involving alcohol consumption often occurred in Kapanara on a single day, and sometimes 20 to 40 cartons or more were consumed in the village the same day.

Villagers also drink beer outside the confines of formal recip- rocal relations. The potential inspirations for such events are al- most limitless, from celebrating being released from jail to the purchasing of a steer. At other times, people simply want to get together and drink.

Sometimes two or more individuals, usually men, pool their money to purchase a carton of beer, though more commonly a person buys one or more cartons himself. People often travel together to purchase alcohol in Kainantu. Drinking sometimes starts in the licensed hotel in town and continues intermittently on the trip home on a PMV. Other villagers back in Kapanara eagerly wait to greet those returning to help carry the beer and perhaps receive a few free bottles. Intensive drinking usually be- gins about eight o'clock in the evening and often lasts until dawn the next day. However, the pattern can be quite varied, from short drinking episodes during the day to those lasting several days.

Villagers rarely consume only their own beer. Instead, they give much of it to others and drink beer that they have received. An individual may enter into many transactions involving beverages in a single night, giving two bottles to one person, receiving three bottles from another, and so on. When many people have pur- chased beer, men and, to a lesser extent, women travel from one house to another all night while drinking and exchanging alco- holic beverages. When drinking, the traditional theme of sharing is strongly emphasized, and those who do not share alcohol ac- quire bad reputations. Sharing is revealed not only in giving car- tons and bottles to others but also in pouring a part of the contents of a bottle from which a villager has drunk into the mouth of another.

A party without beer is no party at all. People gather in a house to drink, sing, eat, and exchange stories and jokes. If all goes well, the occasion is extremely enjoyable and one of the highlights of social life. Unfortunately, however, a party rarely lasts the night without a fight. Many fights arise when previous grievances are aired, from anger about the failure to share beer, or from inter- generational or marital conflict. Young men have a bad reputation for starting fights while drunk, and those older men less prone to fighting sometimes request that members of the younger age group stay away from their parties. All the most serious intravillage fights resulting in major physical injuries in 1976 and 1977 occurred during times of drunkenness.

Antagonism between the sexes, a traditional feature of social life, intensifies when drinking occurs. Wife beating is most severe when men are drunk; one husband beat his wife with a hammer, and another broke his wife's arm while intoxicated. Drunk husbands berate their wives for failing to bear them children or for not working hard enough in the gardens. Inebriated men sometimes also scold or hit their lineage "sisters" for failing to give them pork or food on previous occasions. In such circumstances, the "sister's" husband is shamed because of the implication that he has not helped his affine, and he then kills and gives him a pig. This pattern accounts for most of the "drunk-related" killings of pigs noted in Table 5.6.

Personal injuries also increase with drinking. Inebriates are more prone to injure themselves by being badly burned, cutting their feet on open tin cans or on broken beer bottles, or by falling into a drainage ditch.

Beer drinking, as practiced in Kapanara, is often self-perpetuating because incidents occurring during alcohol consumption lead to still more drinking. If an individual passed out from drinking too much and was carried home, those assisting him are later given free beer. If an inebriated man berated his "sister" for not helping him, his shamed affine will give him pork and beer. Individuals who previously were injured while drunk give a party, which includes beer, to celebrate recovery. A drunk man who beats his wife gives his affines compensation, which consists partly of beer.

As reported elsewhere in Papua New Guinea (e.g. Ogan 1966), drinking is associated with atypical behavior. Men, who normally converse in Tairora, speak in *Tok Pisin* and to a lesser extent in *Motu*[12] and the Gadsup language when drinking. Some heavily intoxicated men bite their arms, rip their shirts or the shirts of others with their hands or teeth, cry spontaneously, walk around in a stupor seemingly oblivious to the presence of others, kiss male kinsmen, and, less frequently, chew soil.

When a man spends too much of his household income on alcohol, injures himself, or grossly embarrasses people while intoxicated, others such as his wife, agnates, or cognates may prohibit him from drinking alcoholic beverages for a specified period,

[12] *Motu* is the *lingua franca* of much of the southern half of Papua New Guinea.

usually six to twelve months. Those imposing the prohibition act either on their own initiative or on the request of the drinker. They sometimes plant a shrub (*Cordyline* sp.) to symbolize the prohibition. However, before the prohibition takes effect, those imposing it provide alcohol to the individual for one last drinking bout before abstinence. Inevitably, abstinence during the entire designated period is rare; the man usually begins drinking again after a few months because someone else insists on sharing a drink with him. When the prohibition is violated, the individual often is required to pay the person initiating the prohibition a small cash penalty, usually K5.00 or less. However, no one seems to be upset that the individual did not abstain for the prescribed period. Once the prohibition has been violated, the individual continues to consume alcohol. To celebrate the end of his prohibition, those who imposed it provide a party at which the man is given both pork and beer; at this time, the spirits of the ancestors sometimes are requested to watch over the person to ensure that he does not injure himself or get sick while drinking alcoholic beverages.

Men and women voluntarily abstain from drinking on certain occasions. After the death of a close kinsman, people traditionally refrain from eating one or two foods such as yam or pork for about six months or longer as a sign of mourning; some now abstain from beer as well. Women in the later stages of pregnancy and mothers actively nursing an infant do not drink, fearing that alcohol consumption will injure their child or that their breast milk will dry up.

Although several negative aspects of alcohol consumption have been described, from the perspective of many villagers the activity is nevertheless one of the most enjoyable available. Kapanarans themselves give several reasons for drinking. One is that they enjoy the feeling of being drunk and the humor involved. Another is that during funerals drinking enables them to cry with less inhibition. Certainly, the ease with which the "cult of beer" has been integrated with the exchange system is another important reason for its popularity. Communal values of sharing and reciprocity, important aspects of traditional society, continually are expressed and reinforced. Another characteristic of traditional society, competition, also is evident in alcohol use. Several Kapanarans are proud of their ability to consume large amounts of beer; one noted that whereas people from other villages become inebriated after consuming four or five bottles, Kapanarans can drink much more. Certainly, drinking fills a great void in social life caused by the cessation of intervillage warfare and the decline of many traditionally meaningful rituals and customs.

The powerful lure of alcohol creates a greater demand for money and hence strengthens the commitment to commodity production and other cash-earning activities. Because commodity production and subsistence endeavors are closely linked (see below), beer drinking also exerts a strong influence on the subsistence system.

CARD PLAYING

One of the favorite pastimes of the people of Kapanara is playing cards, or *laki* in *Tok Pisin* (see Laycock 1972: 476-477). Villagers play cards mostly during the day at the hamlets or along the roadsides and to a lesser extent at night in homes lit by a kerosene lamp. The most popular games are called "three leaf," "seven play," "queen," and "last card." Usually five to ten individuals sitting in a circle are involved in the activity, with several others watching intently. On some occasions, I have seen more than 20 people gathered together either participating or observing. Often more than one game is occurring in different areas of the hamlets. Adults and children of both sexes enjoy the activity, though adult men play cards most frequently. Men usually gamble with other men, but games involving both sexes can also be found.

Men play largely for money. The size of the stakes depends on the time of year. During the coffee flush, when more money is available, men not only play cards more frequently, but the size of the stakes also increases. At this time, they often win or lose K10 to K20 during a single day of card playing. If a player loses a very large amount, such as K50 or more, the winner sometimes gives him back K5 or K10. In a typical game, men first make a bet and then receive their cards. When several gamblers are involved, there are often two or more separate pots. If the winner placed money in all the pots, he gets all the money, but if he bet in only one pot, he wins that one only. The person with the next best hand participating in the other pots gets the remaining money.

Although men occasionally gamble with people from other villages, most money involved circulates within Kapanara. Some people are chronic losers, whereas others seem to be successful most of the time. Without further study, it is impossible to assess whether gambling helps redistribute wealth within the village, though some informants suggested that a few chronic losers were also those with low incomes. Wives complain bitterly when their husbands lose relatively large sums, and some men do not return home until their wives' anger has partially abated. Even with such obstacles, people remain confirmed gamblers. The intense desire to play cards for money, like the emphasis on drinking beer, is a powerful force motivating people to engage in cash-earning activities.

VILLAGE TRADE STORES

In the mid-1960s, a Kapanaran opened the first villager-owned trade store, or *haus kantin* ("canteen") in *Tok Pisin*. Since then, the number of stores in the village grew slowly until 1976 when it reached a high of nineteen, or one store for every six households. The peak coincided with the influx of coffee income due to high producer prices, which dramatically increased village purchasing power.

Those men starting the original trade stores did not have sufficient funds to purchase stock and therefore usually obtained contributions of K10 to K20 from four to six other individuals. Many conflicts later arose over the distribution of store profits, and consequently, most trade stores today are either individually owned or owners have obtained contributions from only one or two others. Many who previously operated a trade store were not doing so in 1976 and 1977. Some ran a store for one or two years but decided not to continue operating because of the annual license fee of K6.00. Others abandoned their business for lack of profits or cash reserves to purchase stock.

Trade stores are smaller than houses, occupying an area of approximately seven square meters. They are constructed largely from traditional materials—bamboo, timber, and grass—though a European influence is evident. Like the European-run stores, they all have a rectangular floor plan, and in 1977 three stores had a corrugated metal roof, a symbol of *bisnis* clearly distinguishing them from other structures. Most owners build their trade store next to their house, both for convenience and to guard against thefts. When an owner changes residence, he often allows a trusted neighbor to manage the store until the building has rotted, at which time he builds one next to his new home.

The items most commonly stocked in trade stores are canned fish, rice, crackers, tobacco, newspaper for rolling cigarettes, soap, kerosene, and matches. Other items, sold in only a few village stores, are canned beef, Cheese Pops, packaged peanuts, cigarettes, airmail envelopes, batteries, flour, grease dripping, clothes, cabbage seeds, face decorating paint, salt, marbles, flavored drink, chewing gum, coconuts, betel nuts, and lime. Inventories of six stores in August 1976 and June 1977, periods when the stores were most fully stocked, revealed that the largest retail value of the items held in a store was K42.35 and the lowest was K17.00, indicating the small scale of the operations.

Because of the relatively small cash requirements for opening a trade store and the increase in coffee prices in the mid-1970s, many adults had the financial resources to start one, and a few individuals with relatively low incomes did own a trade store. Nevertheless, cattle bosses controlled a disproportionate percentage of the stores, running about half of those (or slightly less) in operation at any one time.

Store owners purchase most of their items in bulk in Kainantu stores, except for coconut, betel nut, and lime, which are usually bought in the Kainantu market. More rarely, they travel approximately 80 kilometers to the Markham Valley (Morobe Province) in the lowlands to purchase betel nut to sell at the store. The markup on prices in the village varies greatly but is most commonly between 30 and 60 percent (see Table 6.7).

No one remains in a trade store during the entire day. Only when a family member of the trade store owner or manager is in the vicinity is it possible to purchase goods. Adults often send their children to make purchases. With the little money they

TABLE 6.7

Percentage Markup on Items Sold in a Trade Store

Items	Purchase Price (toea)	Sale Price (toea)	Percentage Markup
Rice (per parcel)	20	30	50
142 g canned fish (per can)	12.5	20	60
425 g canned fish (per can)	28.6	40	40
Tobacco (per pack)	22.6	30	33
Flavored drink (per bottle)	8.3	20	141
Chewing gum (per piece)	2.4	5	108
Matches (per box)	2.3	3.3	43
Soap (per bar)	42.9	60	40
Crackers (per 31.25g)	3.6	5	37

have, children also buy chewing gum, crackers, and Cheese Pops for themselves. Quantitative data comparing the amount of money villagers spent purchasing items in Kapanara with that at stores in Kainantu are not available, but most purchases are definitely made in the village.

Most purchasers pay cash, with store owners actively discouraging customers from asking for credit. Many stores have a sign outside stating in *Tok Pisin*, "*Dinau i tambu hia*," or "loans are forbidden here." Nevertheless, owners do occasionally grant credit, though the amounts are relatively small. They charge no interest on the credit. The low level of Kapanarans' commercial cash debt to fellow villagers is characteristic of the region.

Not all trade stores remain open the entire year. Store operations reflect seasonality in the village cash economy, which is closely tied to coffee production. As more coffee is sold, more money is available in the village to spend at trade stores. After the coffee flush, the purchasing power of villagers drops, and as a result many trade store owners close their business until the next coffee season. The number of stores operating during different times of the year thus parallels seasonality in coffee production (see Figure 6.6).

The volume of store sales also rises and falls with coffee production. Table 6.8 records the sales of two trade stores at different times of the year.[13] September 1976 and July 1977 represent the end and middle of the coffee flush respectively, whereas the periods February and October-November 1977 are characteristic of conditions when coffee income is very low.

The comparison shows that the rise and decline of trade store sales as indicated by both the value and volume of sales follows a trend similar to coffee sales. After the coffee flush, the actual level of all Kapanaran purchases declined more than these figures indicate because the size of the two trade stores' tributary area increased after the coffee-harvesting season as many other stores ceased operation. Thus, even though the two trade stores were serving larger tributary areas after the flush, their sales still declined.

The increase in trade store sales volume during the coffee season is also reflected in dietary data (see Appendix C.2). For example, the consumption of rice and fish, the two most commonly

[13] Daily records of sales, noting the purchaser, the goods bought, and the amount spent, were kept by the owner or a family member. The months or periods listed in Table 6.8 were determined partly by the duration of the records.

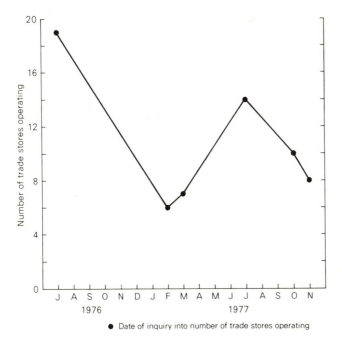

Figure 6.6. Seasonality in Kapanara Trade Store Operations

purchased foods, rises and falls with village coffee income (see Figure 6.7).

Clearly, store-bought foods have become a popular item in the village diet. People serve them on almost all social and ceremonial occasions and consume them as snacks and at household meals. However, without information on the total amounts of foods eaten, it is difficult to assess adequately the nutritional significance of store-bought foods. Certainly, canned fish and meat are important sources of high-quality protein, though many other purchased foods are highly refined carbohydrates that are nutritionally inferior to many home-grown vegetables. Increased consumption of refined carbohydrates can eventually lead to increased dental caries, obesity, chronic degenerative diseases in later life, and nutritional deficiencies such as hypovitaminoses (Jeffries 1979: 202).

CASH-RELATED ACTIVITIES AND SUBSISTENCE PRODUCTION

The scheduling of productive activities is a basic characteristic of all agricultural systems (Flannery 1968; Nietschmann 1973). In scheduling such activities, people are cognizant of seasonality

TABLE 6.8

Seasonality in Trade Store Sales (in Kina)

	Sales of Trade Store A		Sales of Trade Store B	
Item	September 1976	February 1977	July 1977	20 October to 19 November 1977
Canned fish	32.20	11.10	13.90	7.00
White rice	24.40	16.50	13.00	6.30
Biscuits	16.10	4.00	4.80	0
Soap	4.30	2.30	0.60	0.50
Kerosene	2.90	0.40	0	0
Chewing gum	2.05	0	1.40	0
Flavored drink	0.20	0.40	0	0
Matches	0.90	0.50	0	0
Tobacco	11.10	9.80	18.60	11.70
Newspaper	1.50	1.00	0.70	1.30
Cheese Pops	0	0	4.10	0
Peanuts	0	0	2.70	0
Canned beef	0	0	0.80	0
Total Sales	95.65	46.00	60.60	26.80
Value of sales per day	3.19	1.64	1.95	0.86
Number of sales*	246	98	172	69
Percentage of total sales that are dietary items	78	70	67	50

* If an individual purchased items more than once in a single day, each visit is counted as one sale.

in weather patterns and in changes in the availability of biotic components in the local environment. Commodity production and other externally derived endeavors such as card playing and beer drinking must be integrated into the pre-existing scheduling framework. The gross amount of time spent on these introduced activities as well as seasonality in scheduling both leisure and productive activities heavily influence the structure and resilience of subsistence systems.

In Kapanara, most income-producing effort takes place during

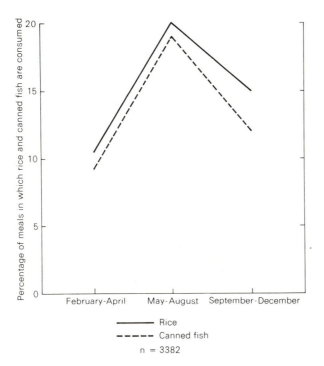

Figure 6.7. Seasonality in the Consumption of Rice and Canned Fish, Sample of Households

the coffee flush. Therefore, to understand the impact of commodity production, it is first necessary to examine the subsistence activities appropriate to the dry season, which occurs during the flush from May to August. A critical issue is the extent to which subsistence gardening activities are constrained by the dry season. If the constraints are considerable, with little subsistence work possible during the dry season, the increase in commodity production would not compete seriously with previously established gardening patterns. The converse is also true.

Considerable flexibility is possible in scheduling inputs into the village subsistence system because gardeners utilize a variety of environmental zones in which they plant numerous crop combinations with different tolerances of dryness. For example, people plant sweet potato in gardens in the moist-to-swampy soils in the valley bottom throughout the dry season because the water table there is sufficiently high to ensure adequate crop growth. Villagers also plant the tuber in the short, dry grasslands (*baeha kai'a*) and in the forests (*kaati kai'a*) in July and August. In addition, through-out the year in most environmental zones, people replant sweet potato gardens after removing and burning the old vines.

Not only is sweet potato planting possible during the dry season, but a host of other tasks can also be done in preparation for later planting from September to December. To plant taro and taro kongkong gardens in rain forest, gardeners must fell vegetation sufficiently far in advance of the onset of heavy rains to ensure that it dries adequately to facilitate burning. Traditionally, this activity began in June through August. During the dry months, villagers also clear grassland vegetation and dig drainage ditches to prepare for later planting.

Thus, the dry season poses no major environmental constraint to preparing and planting new gardens, especially because Kapanara has a large area with moist-to-swampy soils. Du Toit (1974: 168) also reported a lack of seasonality in sweet potato gardening among a group of nearby Gadsup people:[14] "There is no particular time of the year when gardens are started or specific seasons when they believe it to be most advantageous for them to plant their products." However, even though sweet potato cultivation in Kapanara is not markedly seasonal, the planting of taro and taro kongkong largely from September to December gives the overall village planting pattern a slight bias towards the end of the year.

To compare the amount of time that married adults in the sample spent on subsistence production with their inputs into commodity production and other introduced activities, I use data from a time-allocation study I conducted in 1977.[15] In Figure 6.8, the time they spent on clearing, burning, drainage ditch construction, tilling the soil, and planting is aggregated to reveal the marked seasonality of inputs into creating new gardens. Their inputs into coffee production in 1977 were also seasonal but followed the opposite pattern (see Figure 6.9). The relationship between coffee production on the one hand and beer drinking and card playing on the other has been established. Seasonality in the time spent on the latter two activities (Figure 6.10) parallels that of coffee

---

[14] Du Toit (1974: 168) also hinted at a tendency for the Gadsup to plant in November, but it is not clear whether he was referring to forest gardens only or to all types of gardens.

[15] To study the allocation of time by members of the sample, I used a modified version of the time-allocation study method devised by Johnson (1975). A detailed presentation of all the data is in Appendix C.3. For more information on the methodology employed, see Grossman (1979, 1984).

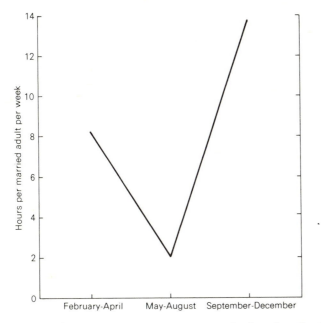

Figure 6.8. Seasonality in Time Spent Preparing New Gardens, Sample of Households

production. "Card playing" includes both gambling and playing without money. Married adult men spent on average 13.5 hours per week playing cards during the coffee season, almost twice as much time as they spent on subsistence-gardening activities during the flush. "Beer-related activities" include such actions as drinking and carrying cartons of beer. However, less obvious instances of beer-related activities, such as an individual being idle because of inebriation, were not counted because of the subjectivity involved in the determination. In addition, most beer drinking occurred at night when data on time allocation were not collected. Thus, the extent of beer-related activities is under-represented in the figure. Wage labor outside the village also varied during the year, with most inputs occurring from May to August when people picked coffee at the plantation at Noraikora (see Figure 6.11). In Figure 6.12, the seasonality of inputs into preparing new gardens is compared with the amount of time allocated to coffee production, wage labor outside the village, beer drinking, and card playing.

The trends in Figure 6.12 clearly suggest that labor inputs into creating new gardens were limited by the allocation of time to

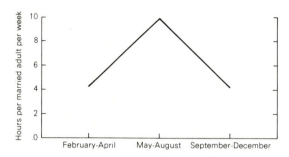

Figure 6.9. Seasonality in Time Spent on Coffee Production, Sample of Households

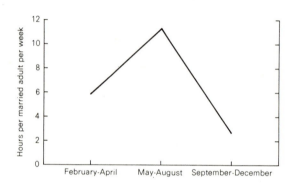

Figure 6.10. Seasonality in Time Spent on Card Playing and on Beer-Related Activities, Sample of Households

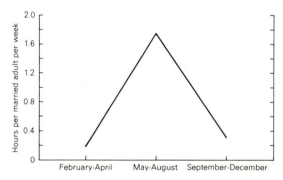

Figure 6.11. Seasonality in Time Spent on Wage Labor outside the Village, Sample of Households

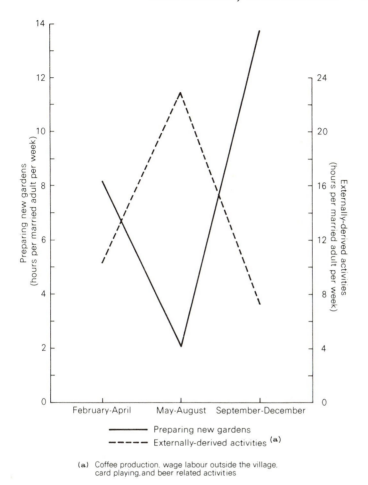

Figure 6.12. Seasonality in Time Spent on Preparing New Gardens and on Externally Derived Activities Compared, Sample of Households

commodity production and other cash-related activities in 1977.[16] The trends may have been even further exaggerated in 1976, when more people picked coffee during the flush at the nearby plantation and the coffee trees in Kapanara carried a heavier crop. Although some seasonality in the effort to prepare new gardens

[16] Although a comparison of these data with time-allocation studies from other areas of the Highlands (see Waddell 1972: 97; Howlett *et al.* 1976: 210-214) might appear useful, it is fraught with many difficulties. Data collecting methods, the consumer-worker ratios of the samples studied, environmental parameters, and

must be expected, the degree of seasonality revealed in the figure could not be due to environmental limitations alone. Villagers themselves recognize the conflicts between subsistence and the commercial economy. One man observed that other people should have been clearing forest in August to prepare new gardens but were not doing so because, he said in *Tok Pisin*, "*kopi kalabusim mipela*" or "coffee 'imprisons' us."

During the year, married adults in the sample spent on average only 1.7 hours per week working on the cattle projects compared with 6.3 hours per week on coffee production. They expended relatively little effort maintaining the barbed wire boundary fences, planting improved pasture, providing veterinary services, and handling cattle. However, as is usually the case, their labor inputs into cattle raising in 1977 were concentrated into the period at the end of the coffee flush (see Figure 6.13). During this time, villagers cleared grass beneath the fences to prevent the wooden posts from rotting and to create a firebreak. They also constructed new subdivisions and fence extensions. Thus, scheduling a large portion of the inputs into cattle raising at the end of the flush accentuated the conflict between subsistence and cash-related activities revealed in Figure 6.12.

The large amount of time devoted to commodity production, wage labor outside the village, card playing, and beer drinking increased seasonality in subsistence production and hence in the supply of locally produced food. Thus, married adults prepared most new gardens from September to December, a seasonal pattern clearly reflected in the consumption of maize, which ripens approximately four months after planting (see Figure 6.14). In addition to causing increased seasonality in food production, the allocation of considerable time to income-related activities during the flush delays forest clearing, thus making the subsistence system more vulnerable to unseasonal weather patterns, particularly an early wet season.

Gardening was not the only subsistence activity hindered during the 1977 coffee season. The quality of pig husbandry also

population densities all differ and may account for variations in the recorded labor inputs into subsistence activities. In addition, special circumstances can produce a pattern different from the normal routine. For example, the time-allocation study reported by Howlett *et al.* (1976: 210-214) was conducted during a drought, and Waddell (1972: 93) collected his labor input data partly during a period of mourning. Both factors may have reduced the normal level of inputs into the subsistence systems.

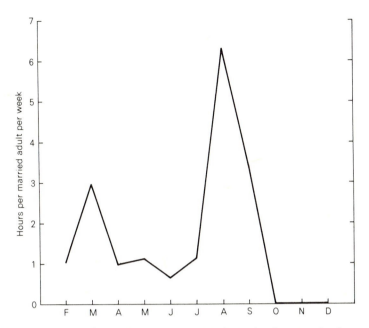

Figure 6.13. Seasonality in Time Spent on Cattle Husbandry, Sample of Households

declined. During the flush, villagers fed their pigs less frequently and gave them less food than at other times of the year (see Figure 6.15).[17] Occasionally, people came home late from picking coffee and thus did not have enough time to harvest tubers for pig rations, making the animals hungrier and more determined to break into cultivated areas. Extensive gambling and beer drinking also resulted in less time spent on pig husbandry and increased the likelihood of garden invasions. After drinking throughout the night, men and women rarely engaged in productive activity the next day. Some women, for example, were too tired to harvest for their

[17] The data in Figure 6.15 were obtained during one-week surveys. The information had to be adjusted because of variations in the size of the pigs fed during the three observation periods, as the quantity of rations provided to pigs depends partly on their size. To determine the adjustment necessary for each period, the number of pigs in each size class (small, medium, and large) was multiplied by the appropriate consumer unit (0.2, 0.3, and 0.5 respectively) and the results summed to yield the weighted average. The figure for March is the actual amount fed. To obtain the figures for July and November, the weighted average for March was divided by those for July and November respectively and the result multiplied by the actual amount fed to yield the adjusted weight of the rations.

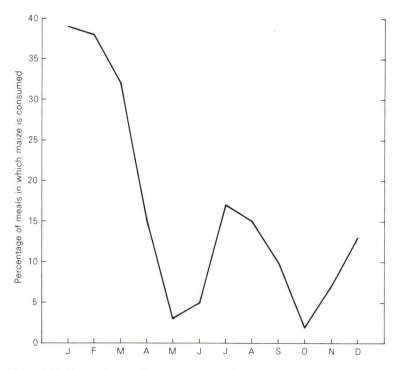

Figure 6.14. Seasonality in the Consumption of Maize, Sample of Households

pigs. In addition, men devoted the most time to card playing and beer drinking, but they were also the ones responsible for garden fence maintenance to keep out the pigs. They ignored the fences making it easier for pigs to break into and destroy gardens. Consequently, complaints about pig damage to gardens became more frequent during the flush.[18] One woman who shared a common garden fence with several men chastised them for failing to maintain the fence, declaring

> You think only of beer. I asked you to fix the fence, but you do not listen. Food is the most important thing. Beer is only water. Food supports your stomach. You do not think of the garden. The pigs have spoiled the garden and have broken the fence here.

[18] The increase in the time that men spent on pig husbandry from May to August noted in Table C.1 of Appendix C.3 reflects an increase not in feeding pigs, but in the hunting of both domestic and feral pigs, some of which spoiled gardens.

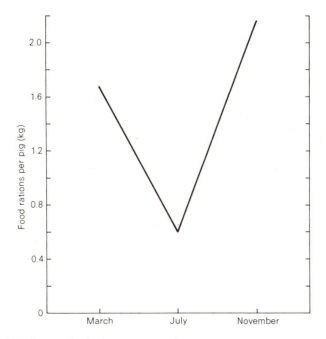

Figure 6.15. Seasonality in the Provision of Pig Rations

The powerful influence of gambling and beer drinking indicates clearly that the way people spend their money also affects subsistence production. Indeed, the more money earned, the greater the likelihood of increased beer drinking and gambling, at least up to a certain point. The village magistrate recognized this relationship when he pleaded with others not to engage in additional wage labor at the nearby coffee plantation because he felt that they would only spend their earnings on beer. Thus, commodity production and other cash-earning activities can have an additional impact on subsistence production because the money earned is used to support activities such as gambling and beer drinking that are also detrimental to subsistence. The cash sphere is also self-reinforcing. If villagers come home too late to harvest food after picking coffee, they can purchase food at the trade stores. Although in the short run the villagers may find buying food a satisfactory alternative to relying on subsistence production, the resulting reduction of local food output the following year caused by a decline in preparing new gardens could have serious consequences. If at that time commodity prices were to fall markedly or if unfavorable weather further limited local food production, food shortages could result (see Chapter Seven).

From Subsistence Affluence to Subsistence Malaise

In the mid-1970s, villagers still viewed subsistence production as an integral part of their life but clearly valued commodity production such as cattle raising and coffee growing more. This disparity in evaluation was reflected in many of their statements:

> Before we planted gardens at all times [of the year]. Now *bisnis* has arrived, and we think only of *bisnis*.

> Before, we thought only of gardens and building strong garden fences. Now, this time, everyone plays cards, forgets about the fences which rot, drinks beer all the time, and is lazy; people do not care for gardens either. They are bored with gardening. If they are hungry, it is their own fault.

In a dispute over the use of land within the confines of a cattle project, a magistrate from another village criticized the Kapanarans for their overenthusiastic commitment of land to income-producing ventures:

> Later, your children and grandchildren will be born, and there will be many people here. Where will they live and cultivate gardens? Where is there room for them to live? . . . There will not be enough land. You think only of *bisnis*. . . . You have used all the land for commercial activities. You are only happy with *bisnis*. You must leave some ground [for gardening and other uses].

Although I cannot assert that the higher valuation of commodity production is universal in Papua New Guinea, I have found evidence for it in other villages in the Eastern Highlands Province. An item in a newspaper (PNG, *Post-Courier*, April 10, 1979, p. 15) also reported a similar condition in the Western Highlands Province. Members of the Western Highlands Chamber of Commerce were concerned that restrictions on rice imports might lead to a food shortage, particularly during the coffee season because:

> the spread of cash cropping had resulted in people neglecting their food gardens. . . . good coffee seasons had brought a reliance on cash cropping, and the women were now reluctant to return to traditional patterns.

As a result of the intense enthusiasm for commodity production and the commercial economy in the mid-1970s, subsistence affluence, which had been a distinguishing characteristic of the Kapanaran production system since pacification, had been replaced by what I call *subsistence malaise*. This condition developed when the previously strong commitment to subsistence production declined because of a negative comparison with commodity production and other externally derived activities. It resulted in a reduction in local food output and a deterioration in the resilience of the subsistence system.

CHAPTER SEVEN

The Bust: The Decline of the Commercial Economy in the Early 1980s

IF THE RAPID PACE of change occurring during the boom of the mid-1970s had continued, major transformations in the village production process would likely have resulted, leading the Kapanarans on paths followed by many other peasant communities in the Third World. However, the boom proved to be ephemeral. By the early 1980s, Kapanaran involvement in the commercial economy had declined radically because of a dramatic drop in the price of coffee, a reduction in agricultural extension assistance for smallholder cattle projects, and problems within the village. Because Kapanarans retained considerable autonomy, they were able to reverse at least partially their commitment to commodity production and the commercial economy. Consequently, many problems associated with cash-earning activities also declined, though certain stresses on subsistence production remained.

CHANGES IN SETTLEMENT AND SUBSISTENCE

Upon returning for a brief visit to attend a ceremony in Kapanara in early April 1981, I was struck immediately by the major transformation in settlement pattern. Although nucleated hamlets remained near the main road and the coffee groves, many Kapanarans—approximately one-third or more of the households—were not living in them. Instead, they had built new houses in groups of two or three scattered far from the nucleated hamlets, either at the forest-grassland border or in the grasslands near the forests. A few had done so in late 1980 and many more in early 1981. People traditionally build temporary shelters at remote garden sites, but the new structures I found in 1981 were properly built houses, similar to those in the hamlets. This alteration in settlement pattern was linked to changes in subsistence and commodity production.

The villagers experienced a marked food shortage from about August 1980 until May 1981, during which supplies of sweet potato were especially low. Kapanarans asserted that in order to produce food as rapidly as possible to alleviate the food shortage they cultivated in areas far from the hamlets. By preparing new gardens in remote areas in or near the forests they could save time and energy by not having to construct garden fences, because the likelihood of pigs roaming in such places was, they hoped, minimal; having to construct fences—a labor input necessary in gardening closer to the hamlets—would only prolong the hunger period. Similarly, their decision to live near the new food gardens they were preparing was motivated by the need to produce food as quickly as possible. By residing permanently near their new garden plots, they would not have to walk the long distance to and from the hamlets each day. Instead of having to rest after arriving at the plot, they could start work right away early in the morning; nor would they have to stop work and leave early to return to the nucleated hamlets. They would also avoid being drenched by rain while travelling to their plots, soakings that normally can weaken people's resolve to work hard. Interpersonal competition also played a role. Some worried that if they remained at the nucleated hamlets, they could not produce the needed food supplies as rapidly as those living near their remote gardens, and wishing to avoid invidious comparison, they also moved close to their new gardens.

The decision to live near the forests did not signify that in areas far from the nucleated hamlets Kapanarans no longer feared attacks of sorcery or *sangguma* from people in other villages. Rather, it reflected their overriding concern to produce enough food *in spite of* the fear of attack. People built their new houses in clusters of two and three units partly to provide security, believing that men from other villages would less likely attack in areas where other Kapanarans were also residing.

Although the Kapanarans' explanations for the settlement changes were informative, something about the situation was still puzzling. This was certainly not the first food shortage they had experienced. Indeed, the dearth of food supplies from late 1975 until the end of 1976 was equally severe, if not worse, and that problem did not result in a major change in settlement pattern. So why was there a different settlement pattern in 1981?

The answer in part reflects the continuing pressure of commodity production on subsistence activities. In 1976, villagers were still preparing new gardens near the nucleated hamlets. Even though the depredations by pigs were extensive, people still perceived the unenclosed area near the hamlets as valuable for gardening, and consequently they remained in the nucleated hamlets. In contrast, by 1981 they were cultivating extremely few new plots near the hamlets. Not only were pigs still a problem, but garden destruction by marauding cattle had skyrocketed dramatically from the levels in the mid-1970s. Because of continuing conflicts between cattle bosses and followers over the distribution of project revenues, followers in several projects ceased providing labor regularly, and consequently the barbed wire boundary fences fell into disrepair. Cattle from the projects were thus able to escape from the enclosures; they usually waited until night to roam the village, destroying most food gardens near the hamlets. Some cattle, in fact, wandered into a few forest gardens and devastated them as well. Indeed, the problem was so severe that irate gardeners began shooting arrows at cattle invading their plots. From 1980 to mid-1981, they killed at least 12 head and wounded more.

After the food shortage was over in May 1981, villagers moved back to the nucleated hamlets. As a result of preparing most new gardens in remote areas, the spatial distribution of subsistence plots in relation to the hamlets was now even more skewed than it had been in the mid-1970s. Distances to cultivated areas had increased even farther.

Kapanaran explanations of the cause of the food shortage varied. Garden destruction by cattle, of course, was obvious to everyone. For some, sorcery was also to blame. Men from the Auyana area to the southwest were outraged because after the death of a female relative, the deceased's Kapanaran husband gave them a very small mortuary payment. Dissatisfied, the Auyana men supposedly cast a spell stunting the growth of sweet potato.

Many Kapanarans, however, felt that the hunger was their own fault. One asserted, "People were drinking too much beer and spending too much time playing cards. They became lazy and just 'slept' at the hamlets [i.e. did not work hard in their gardens]." Apparently, villagers did not prepare as many new gardens as they should have. Several also reported that instead of households having four to eight producing garden sites in different parts of the village territory as they had had previously, many gardeners in 1980 had only a few plots, reflecting their desire to reduce labor inputs into subsistence production. Although they made these new plots larger than the more numerous scattered ones cultivated before, people were also more vulnerable to total destruction of their food producing areas by marauding cattle and pigs. Whereas previously the loss of food output resulting from an invasion of one of several gardens would not have been overly disastrous, now the destruction of one of only a few gardens had much more devastating consequences. Such reports indicated that the condition of "subsistence malaise" was still current.

Several years of good coffee prices had lulled people into a false sense of security, making them complacent about subsistence production. Even though coffee prices had dropped from the 1977 high levels, prices during the flushes in 1978 and 1979 and in the early part of 1980 were still favorable compared to the early 1970s (see Figures 7.1 and 6.1).[1] Villagers felt that with good coffee incomes continuing, they could purchase the food they needed if subsistence production declined. However, starting in August 1980 the price of coffee plummeted. For example, in the first week of July 1980 the average factory door price of dried coffee was K1.27 per kilogram, but in the first week of August it had dropped to K1.04 per kilogram, and by the end of the year stood at only 78t per kilogram. Along with coffee prices, village coffee income also fell, leaving most people with little to fall back on as their food supplies were also low. A few of the wealthier individuals relied heavily on store-bought foods during the food shortage, but many Kapanarans were hungry and some had to obtain gifts of food from relatives in nearby villages to ameliorate their situation.

Intensive inputs into coffee production during the first part of the 1980 flush also affected the food-supply situation. The coffee trees bore a heavy crop in 1980. Thus, in the early part of the coffee season while prices remained favorable, villagers spent much time picking all the available berries, believing that the year would be an excellent one financially. Consequently, their inputs into preparing new food gardens in May, June, and July were depressed, similar to the pattern observed in 1977. Hence, the extent of the food shortage was exacerbated because sweet potato gardens that could have been started then would have been ready to yield tubers beginning around December 1980. However, the food shortage persisted until May 1981, because Kapanarans did not start making intensive inputs into preparing new food gardens until August 1980, when the price of coffee dropped. Because of the low prices, after August they made minimal inputs into coffee

[1] However, the full impact of the price increase in 1979 was not felt because of poor coffee yields that year.

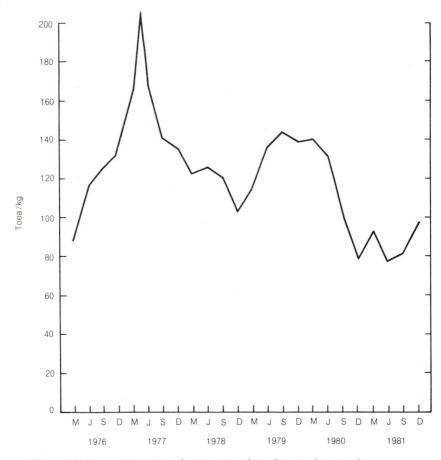

Figure 7.1. Factory Door Price for Dry, Sound Quality, Arabica Parchment Coffee, 1976-1981. *Source*: Papua New Guinea Coffee Industry Board, Goroka.

production, harvesting only a portion of the crop, paying little attention to weeding, and planting few additional trees.

Pigs also contributed to the problem, just as they did during the last food shortage in 1975 and 1976.[2] Once again, they received fewer sweet potato rations, making them hungrier and more prone to break into fenced gardens and to go farther afield to invade unfenced ones. Complaints about pig damage to gardens increased during the food shortage. However, the severity of the pig problem subsequently decreased as food production rose by May 1981.

[2] I estimate that the size of the pig herd was approximately 220 to 250 during the food shortage. Thus, the herd was not abnormally large.

Rainfall patterns in 1980 also undermined the food supply. An agronomist at the Highlands Agricultural Experiment Station at Aiyura (M. Bourke, personal communication 1981) noted that many people in the region reported reduced sweet potato yields in late 1980 and early 1981; the erratic rainfall pattern in 1980— characterized by alternating very wet and dry months—may have caused the problem. Also, sweet potato normally requires a relatively dry period during the latter stages of growth to produce adequate tuber yields. However, abnormally heavy rainfall in June 1980 (210 millimeters) most likely hindered tuber development in the crop planted in early 1980. Villagers complained that the sweet potato vines in their gardens grew well but the tubers produced in the mounds were smaller than usual. Excessive rainfall during the latter stages of growth can produce such a condition (M. Cannon, personal communication 1983).

The events leading to the food shortage highlight a widespread problem associated with increasing commodity production. Where peasants are involved extensively in the commercial economy at the expense of subsistence production, community welfare in general and local food output in particular often become more vulnerable to unfavorable local environmental conditions such as irregular rainfall or drought (see also Nietschmann 1979; Shenton and Watts 1979). That is, the *resilience* of the subsistence system—its ability to withstand "shocks"—declines. The Kapanaran commitment to the commercial sector, beer drinking and gambling, subsistence malaise, and the related problem of cattle and pigs destroying gardens all hindered food output and distorted the traditional timing of subsistence labor inputs. Thus, when unfavorable weather further reduced food production, the combined negative forces affecting subsistence caused the food shortage. If the Kapanaran commitment to the commercial economy (and its associated problems) had been significantly less intense, the weather pattern in 1980 would not have had such an impact. However, because of the various pressures on the subsistence system, a normally minor jolt to the system became a severe shock with more far-reaching consequences.

CHANGES IN OTHER CASH-RELATED ACTIVITIES

Marketing food in Kainantu is generally undertaken to supplement the village income when coffee revenues are low, but this activity was also depressed in 1981. Obviously the food shortage was to blame in the first third of the year, though sales of nuts from *Pandanus* and *Areca* sp. palm trees did provide some income at this time. Marketing could have increased substantially as local food output rose in May, but it did not because Kapanarans feared sorcery attacks either on the way to Kainantu or in the town itself; people from a nearby village accused them of performing sorcery on one of their members, and expecting revenge, the Kapanarans were hesitant to pass through that village on the way to town. Consequently, marketing revenues remained low during the year.

Nor did wage labor compensate for the low village income. Although several adult men worked as road laborers for the Public Works Department, very few sought wage labor at the nearby coffee plantation at Noraikora. Wages there, like the price of coffee, were low, and the Kapanaran fear of being attacked by sorcery while working there remained strong.

With income from coffee, marketing, and wage labor remaining depressed in 1981, commercially related activities in the village also declined. Trade store owners reported that sales during the 1981 flush were lower than those in 1977. Beer drinking was also affected, with villagers purchasing much less compared to the large quantities consumed in the mid-1970s.

Kapanarans had not lost their enthusiasm for beer drinking— they simply did not have as much money to spend on alcohol. They asserted that when coffee incomes increased in the future, they would start consuming more alcohol again. Beer also became more expensive. A carton of 24 bottles cost up to K12.00 in mid-1981, higher than K9.10 at the end of 1977. However, changes in the way beer was sold from licensed outlets did make it easier to purchase. By 1981, beer was also available in cartons of 6 and 12 bottles. Nevertheless, cases of 24 remained the most popular and prestigious to give in exchanges, though cartons of 12 were also acceptable. Villagers felt that packs of 6 bottles were too small to give in exchanges, because the recipient would be slighted at such an insignificant gift.

The decline in beer drinking was also related to the establishment of a Seventh-Day Adventist church in Kapanara in 1980. A national from another village in the region conducted church services weekly and actively attempted to spread his message to win converts. One prerequisite for baptism in this church is complete abstinence from consuming alcohol.[3] People from about 15 Kapanaran households expressed interest in becoming baptized, and many of them did refrain from indulging in alcohol. Although Kapanaran interest in the church was considerable in 1981, it will likely moderate in the future if past commitments to organized religion are any indication; villagers have been involved previously in several religious organizations—Lutheran, Seventh-Day Adventist, and Swiss Mission—but their enthusiasm usually waned after a few years. Consequently, the level of coffee income will continue to be a more important influence on alcohol consumption than religious affiliation.

The reduction in time spent on coffee production and the decrease in alcohol consumption[4] enabled villagers to increase substantially the time devoted to preparing new food gardens during the 1981 coffee flush compared to the 1977 coffee season. Many men had already started felling trees in forest gardens in June and July, whereas few had done so during the 1977 flush when inputs into coffee production and beer drinking were much greater. The contrasting conditions in the two years support my assertions that commodity production and other cash-related activities conflicted with subsistence production and that the precipitous drop in preparing new food gardens during the 1977 flush was not principally the result of seasonal, local environmental limitations.

Similarly, the impact of commodity production and other cash-related activities on pig husbandry also changed. During the 1981 flush, the quality of pig husbandry improved, and complaints about porcine-induced garden damage decreased compared to the flushes of 1976 and 1977 when the time allocated to externally derived activities was much greater.

The decline in Kapanaran commercial involvement in the early 1980s did not signify the demise of the *bisnis* ethos. Villagers still

[3] Another prerequisite is a ban on pork consumption, but this had only a minimal effect on pig husbandry. Joining a religious organization is an individual, not a household decision. Thus, if a husband decides to become a member, his wife will not necessarily do so. In several cases, only the husband wanted to be baptized, enabling the wife to continue raising pigs. Also, some women who wanted to become church members felt that the prohibition applied only to pork consumption, not to pig rearing.

[4] It was impossible to discover whether time spent on card playing had also decreased, because people can play without gambling or with smaller stakes.

highly valued cash-earning endeavors, but they were not willing to participate intensively in coffee production because the cash returns were so poor and the specter of continuing food shortages necessitated increased efforts at subsistence production. Nevertheless, in August 1981, people enthusiastically greeted rumors that the price of coffee would rise dramatically in March 1982, a reaction reflecting the continuing strength of the *bisnis* ethos.

The Kapanarans were fortunate in being able to reverse their substantial commitment to the commercial economy and to increase subsistence inputs during the bust. They could make the change because their resource base had not deteriorated significantly, they were not heavily indebted to other social classes, and government policies and taxes were not oppressive. Kapanarans' cash requirements were relatively low. The head tax in 1981 remained only K6.00 per male adult, and many people who refused to pay were not penalized by the village court. School fees were only K5.00 per child. Cash was still required for social and ceremonial activities, but the expected outlays to fulfill such obligations were not onerous. Even when hunger threatened, people did not have to intensify cash-earning activities because the continuing strength of the reciprocal exchange system enabled Kapanarans to obtain food from relatives in nearby villages. Without having to increase commercial involvement to purchase food, they could devote more time to subsistence production.

In addition to influencing beer drinking, subsistence gardening, and pig husbandry, the decline in coffee prices also affected Kapanara's terms of trade. The price decline depicted in Figure 7.1 does not reveal the true magnitude of the change because it represents only the average price paid by the processing factories, whereas most villagers sell directly to roadside coffee buyers, not to the factories. In calculating the amount they will offer villagers, roadside buyers must include their transportation costs. The price of gasoline rose markedly in 1979 and 1980, and the buyers had to pass at least some of their own higher transportation costs on to the villagers in the form of lower roadside prices. Consequently, those selling in the village experienced a steeper price drop than that portrayed in Figure 7.1. At the same time, the cost of frequently purchased items had gone up. From mid-1976 to mid-1981[5] the cost of canned mackerel, rice, and tobacco in Kainantu stores rose 60, 17, and 15 percent, respectively, and in the village

[5] These dates are determined by the length of my records.

trade stores, the price of canned mackerel and tobacco[6] increased 37 and 33 percent, respectively. Similarly, the price of beer was 40 percent higher in mid-1981 compared to January 1977. Thus, Kapanarans were earning less while facing rising prices. For peasants bound to the cash economy, such changes are a serious threat to community welfare—part of the simple reproduction squeeze— but because the Kapanarans could reduce their dependence on the commercial economy while increasing subsistence output, the immediate impact of the decline in terms of trade was muted.

SMALLHOLDER CATTLE PROJECTS

Whereas the decrease in coffee production was a favorable development for the subsistence system, the immediate impact of reduced inputs into cattle raising in the early 1980s was not. As cattle were able to escape from the inadequately maintained enclosures, they destroyed many gardens near the hamlets, forcing people to cultivate more land even farther from the residences than previously. Several changes accounted for the reduction in time allocated to the smallholder projects.

Since 1977, the bosses had distributed very little additional revenue to their followers, keeping most proceeds retained in the village for themselves and making the helpers even more bitter. Rumors that some project leaders were secretly hoarding cattle income in their passbooks persisted, and some followers claimed that the bosses' wives did little garden work because the women supposedly purchased large amounts of food with money held by the bosses. Thus, relations between bosses and followers had deteriorated even further compared to conditions in the mid-1970s. Followers in five of the seven projects rarely provided any labor at all, and helpers in the other two worked only intermittently. Although work on the projects continued, the effort was considerably less than in the mid-1970s. In the early 1980s, villagers spent some time changing rotten posts, clearing grass from beneath the fence line, and restraining loose wire. The bosses' households provided most of these inputs, though one project leader still relied heavily on hired labor. The total work effort, however, was not sufficient to maintain fully the barbed wire fences, and consequently cattle were able to escape from the enclosures.

---

[6] Because of the numerous methods of packaging and selling rice in the village, I could not make an accurate comparison of price differences over time.

The decline in the level of agricultural extension inputs—a key variable in Kapanara's external, political-economic environment—also diminished the amount of time devoted to cattle raising. In the early and mid-1970s, extension officers constantly gave smallholders advice, provided them with numerous services, and often directed them to make improvements. This high level of extension input was fundamental to the early success and rapid growth of smallholder cattle enterprises, but in the long run it proved to be counterproductive. The extension emphasis was on supervision and regulation—telling people what to do or doing it for them—rather than on education, which would have eventually made the villagers more self-reliant after extension inputs declined (see Grossman 1980: 28-30). When giving directives, providing veterinary services, or telling villagers to sell their cattle, extension officers often did not consult smallholders or explain their actions. In such a situation, villagers developed a psychological dependence on the extension agents, often hesitating to act on their own (McKillop 1976a: 12). Consequently, when extension inputs into smallholder cattle raising were curtailed in the late 1970s and early 1980s,[7] serious problems arose; without continual external pressure to maintain fences, stockyards, and pastures, management standards fell rapidly. Furthermore, the bosses had difficulty in continuing to justify requests for their followers' labor by referring to extension officers' directives because other villagers were aware that the agricultural agents rarely gave such directives anymore.

The decline in villager and extension inputs also contributed to the reduction in the Kapanaran herd size, which fell from 162 in 1977 to 92 in 1981. One project (no. 7) had no animals at all. As cattle wandered through the village and invaded cultivated areas, 12 head were mortally wounded by irate gardeners, and more broke a leg or suffered other injuries and had to be killed. After 1977, several more animals in the heavily overgrazed pasture died, most likely from problems stemming from inadequate food intake, as efforts at planting improved pasture virtually ceased. In addition, in the case of the project without any cattle, the

[7] The extension effort declined for several reasons. Budget cutbacks limited the extension effort. Also, national extension officers preferred to let smallholders take more initiative in operating their enterprises than did the European officers they replaced in mid-1970s. Furthermore, as problems with smallholder projects mounted, the government placed less official emphasis on assisting village cattle projects.

extension officers never delivered its bull; the original loan provided funding for one bull, ten steers, and seven heifers, but the agents brought only five steers and three heifers. Without a bull, the project was doomed to failure as stock were depleted. Extension officers were no longer delivering cattle to projects in the late 1970s and early 1980s, and thus the project never received the remainder of the herd provided for in the loan.

The decline in herd size was also affected by mortuary obligations in Kapanara. In 1979 and 1980, the number of adults that died was unusually high, and about ten head of cattle were killed and beef distributed during funeral ceremonies. Previously, no one in Kapanara had killed cattle for such occasions.

The falling number of stock, in turn, further demoralized the helpers, because any possibility of ever being compensated adequately dwindled along with the herd size. As a result, they restricted their labor inputs into maintaining the projects even more, contributing to a further deterioration in project conditions, which then enabled cattle to escape more frequently, and so forth.

Even though there were fewer animals in the early 1980s, they were doing more damage to gardens as project fences fell into disrepair. Nevertheless, the reduction in herd size did have an important beneficial effect. Fewer cattle in Kapanara meant less pressure on the local environment. Earlier in the mid-1970s, one project was severely overgrazed and by 1979, two more were facing pasture shortages. However, because there were not as many cattle in the early 1980s, there was much less grazing pressure, and consequently more abundant pasture regrowth and reduced runoff and erosion.

Project gross incomes also declined in the early 1980s compared to the mid-1970s. It was obvious to everyone in the village that sales had slumped. Extension officers no longer provided assistance in marketing cattle from 1978 until mid-1981. Also, with the general decline in coffee income and rural purchasing power, people from other villages came less frequently to buy cattle. Nevertheless, cattle remained a potentially lucrative enterprise, with mature steers selling for K300 each. Future increases in cattle project revenues—which will mainly benefit the bosses—will depend on the state of the coffee market.

The relative wealth of the cattle bosses remained evident in various facets of ceremonial life. For example, project leaders continued to dominate the *singsing* market. The only two held since the mid-1970s were sponsored by cattle bosses, each earning a net income of K300, after deducting expenses. Project leaders also benefited from the use of cattle and cash (coming, in part, from project revenues) in prestations. In 1978, one project leader was involved in an exchange with the *kandere* of his two wives' children. At night, he gave them 20 cartons of beer, which he purchased with his own money, and two small slaughtered pigs; the next morning he also killed two large pigs and one large steer for the exchange. The two groups of *kandere* reciprocated by giving the boss K1,100, the largest cash amount ever received in such an exchange in Kapanara. The boss not only obtained cash, but considerable prestige as well.

Cattle bosses not only used their incomes for household consumption and ceremonial exchanges, but also for commercial investment. One was the largest contributor of nine individuals who purchased a public motor vehicle in 1980, and project leaders operated 6 of 14 trade stores functioning during the 1981 flush.[8]

Most importantly, cattle bosses retained control over the herds and the enclosed areas. Followers owned only 5 of 92 head in 1981, a smaller percentage than previously. Project leaders—even the two involved in the enterprise without cattle—were still able to exclude nonmembers from gardening in the paddocks, and one was even successful in forcing one of his own project members to give up plans to cultivate in the enclosure. Thus, although the rural commercial economy in general and the cattle bosses' income in particular declined, rural economic differentiation persisted.

[8] Through correspondence, I have been informed that the Kapanarans purchased another PMV in 1982. A different cattle boss was the main contributor, though I am not sure how many others also invested.

CONCLUSION

> [O]ne of the major means of achieving rural development is to increase production for the market and improve productivity. It is too often assumed that these processes can be grafted on to 'subsistence' production at no cost to domestic consumption or income. (Heyer *et al.* 1981: 5)

FROM THE PERSPECTIVE of government officials and planners, Kapanara was an ideal setting for stimulating increased commodity production. Not only was there a surplus of total land and labor available to allocate to commercial endeavors, but the villagers themselves were keen to expand their cash-earning activities. Furthermore, factors that elsewhere have hindered peasant involvement in commodity production—major local environmental risks and exploitative government policies—were absent in the Kapanaran case.

Nevertheless, this cultural-ecological analysis of the evolving relationship among food gardens, pigs, commodity production, and associated cash-related activities has revealed that subsistence and commodity production are conflicting, not supplementary spheres. Changing patterns of resource use and production at the village level have seriously undermined the subsistence system. The Kapanarans' enthusiastic commitment to *bisnis* reduced the flexibility of the subsistence system, the amount of food it produced, and its resilience in that it has become more susceptible to local environmental disturbances. Cash-earning activities also lulled people into a false sense of security, leading to a condition of subsistence malaise and its attendant consequences. Intensified involvement in the commercial sphere in the future will similarly be made at the expense of subsistence production.

The unfolding story of Kapanara suggests that contrary to the conventional wisdom, even when surplus land and labor are available, commodity production can have severe consequences for subsistence endeavors. To properly assess the potential impact of the commercial sphere, we must utilize a more comprehensive approach that includes a much broader range of local-level variables.[1]

First, the relative location of commercial and subsistence activities must be examined. The increase in the distance between settlement and food gardens caused by locating cash-earning enterprises near residences is a common pattern throughout Melanesia (Brookfield with Hart 1971: 227; Ward and Hau'ofa 1980: 42). In Kapanara, the increase in distance to subsistence gardens meant a greater daily burden in transporting home the harvest, poorer garden maintenance, and the location of more cultivated plots in the zone of primary danger from sorcery. Also, when coffee prices are high, those engaged in coffee production, beer drinking, and gambling during the flush are less likely to make the effort to travel very far to harvest sweet potato for their pigs, thereby encouraging and facilitating invasions of subsistence gardens. Changes in relative location have been significant in other regions as well. In Malaita, Solomon Islands, cattle projects also enclosed garden land, forcing women from coastal villages to go farther inland to cultivate subsistence crops (Quartermain 1980: 270). Such spatial changes can have important implications for the quality of subsistence output. In Western Samoa, for example, cash crops are located on the best agricultural land, sometimes displacing food crops farther away onto less fertile soil (Ward and Hau'ofa 1980: 42), a pattern that can also be observed in relation to coffee production in many parts of the Papua New Guinea Highlands.

Second, the nature of the linkages within the subsistence system and the local ecological system affects the impact of commodity production. In Kapanara, pigs prefer to root in overgrazed pastures, thus exacerbating the erosion problem, which reduces the agricultural value of the land. The linkage between food gardens and pigs was particularly significant. As villagers became more involved in cash-related activities, they fed their pigs less, and the animals in turn destroyed more food gardens. Because of linkages within the local system—such as those among people, pigs, and gardens[2]—the allocation of resources to commodity production and other cash-related activities not only can detract from

[1] Certainly, the nature of the external, political-economic environment also influences the relationship between the two forms of production. Its role is examined in the section "Persistence of the Peasantry" below.

[2] See Brookfield (1968a; 1973) for an analysis of the interaction among people, pigs, and coffee gardens in a Simbu community in the Highlands.

the time available for subsistence endeavors, but can also set in motion a series of events that can seriously reduce the food output from gardens already established. Similarly, the importance of such linkages in the local system is evident elsewhere. In rural Nicaragua, the hunting of jaguars to procure their skins for sale caused a decline in the jaguar population, which then permitted an increase in the population of white-tailed deer, one of the jaguars' prey. As the deer is a garden pest, crop losses increased (Nietschmann 1973: 178). In Mexico, pasture grasses from commercial smallholder cattle enterprises invaded adjacent food gardens, and because the local food crops could not compete with the seeds and vegetative runners from the pastures, those village gardeners affected had to abandon their plots and establish their own cattle pastures. In turn, those gardening next to the new pastures were forced to switch to cattle raising as well. As a result, this "chain reaction" involving ecological linkages magnified pressure on subsistence output (Dewey 1981: 177-178).

Third, seasonality in the demands for labor as well as the total amount of labor available influences the relationship between subsistence and commodity production. Numerous reports from other parts of the Third World indicate seasonal conflicts in scheduling subsistence and commercial activities. For the Miskito Indians of Nicaragua, the income-producing activities of shrimp fishing, turtling, and rice harvesting competed for time with subsistence agriculture so that the seasonality of labor inputs into subsistence gardening were sometimes out of phase with environmental conditions (Nietschmann 1973: 149). Previously, they scheduled their gardening activities to take advantage of optimal periods of dryness and rainfall. In parts of East Africa, marked rainfall seasonality necessitates an early planting in a relatively short period before the onset of the rains. Porter (1981: 229) cites evidence indicating that there is a 5 to 7 percent drop in total maize yield for each day farmers delay planting the crop after the rains begin. Consequently, planting either subsistence or cash crops at the best time results in less than optimum yields for the other. In the Kenya lowlands, for example, the cash crop cotton and subsistence grains both cannot be planted at the most favorable time because of a labor bottleneck. In addition, cotton weeding can conflict with the harvest of the food crop pigeon pea, and the harvest of late planted cotton may compete for labor with the early planting of the second maize crop (Wisner 1977: 209). In Kapanara, seasonal inputs into coffee production and other cash-related activities reduced the time allocated during the flush to feeding pigs and preparing new food gardens. The conflict in Kapanara, however, was not caused by a labor bottleneck as in Africa, because adult married men were able to spend an average of 13.5 hours per week and married women 5.4 hours playing cards during the 1977 coffee flush. Because inputs into planting food gardens became much more seasonal, the village subsistence system was more vulnerable to unfavorable local environmental conditions, a problem that contributed to food shortages in late 1980 and early 1981.

Fourth, the sociocultural context may intensify the conflict between subsistence and income earning. In Kapanara, intergroup competition motivated people to devote a substantial amount of time in a short period to building the last four cattle enclosures, causing a decrease in garden planting and a subsequent increase in pig damage to cultivated plots. At the same time, villagers were reluctant to get rid of their pigs because the animals are highly valued as an integral part of the reciprocal exchange system and a source of income. Establishing seven cattle projects and enclosing a large area with the attendant spatial impact on the agricultural system were also partly due to intergroup competition. The cultural emphasis on the importance of *bisnis* and the resulting condition of subsistence malaise—factors evident in other parts of the Highlands—undermined subsistence production.

Fifth, the way in which income from commodity production is spent affects subsistence. Beer drinking and gambling compete with subsistence production for time. An increase in cash earning means that more money is available to spend on these externally derived activities, thus intensifying the conflict with the subsistence system. In addition, with larger incomes people can purchase more store-bought foods, thereby decreasing the felt need to maintain subsistence output; declining subsistence production, in turn, negatively affects pig husbandry, which feeds back to harm food gardens.

These five fundamental factors, of course, must not be considered in isolation from each other. Indeed, because they are systemically interrelated, the impact of commodity production reverberates throughout the local system. For example, in Kapanara the sociocultural context affected the relative location of subsistence and commercial activities, which, in turn, altered preexisting linkages among people, pigs, and gardens and influenced spending patterns, particularly during the coffee-harvesting season. Thus, the impact of commodity production becomes magnified greatly.

The Kapanaran willingness to allocate resources to the commercial sphere while threatening their own subsistence base contrasts with the pattern observed by many other researchers that peasants generally aim first to satisfy their own household's food requirements and only then to engage in commercial activities (assuming demands from other classes and the state do not dictate otherwise) (e.g. Wolf 1969: xiv; Tosh 1978: 428; Lofchie 1980: 101; Guillet 1981: 6). For example, in his study of Lango agricultural history in Uganda, Tosh (1978: 436) asserts that the cash crop cotton would have been more productive if planted at the beginning of the rainy season, but the Lango wanted to secure the best possible yield for their food staples, at the expense of cotton yields. However, the Kapanarans are certainly not unique in Papua New Guinea in allowing their overenthusiastic response to the commercial sphere to operate to the detriment of subsistence activities. For example, among the Nagovisi of the North Solomons Province cocoa planting is so extensive in some areas that land for subsistence gardening is very scarce (Mitchell 1976: 139-140). Similar emphasis on commodity production to the detriment of subsistence endeavors has also been observed in parts of East Africa (Datoo 1977), Nicaragua (Nietschmann 1973, 1979), and Brazil (Gross and Underwood 1971). Such varying commitments to commercial activities reflect differences in the degree of local autonomy, the nature of the political-economic environment, and the extent of environmental risk.

Although population pressure in the Highlands has not reached the point where people no longer have access to adequate land to meet subsistence requirements, many areas are not as fortunate as Kapanara in having such an abundance of land. Where land for subsistence gardens is currently adequate but land for expanding commodity production is in short supply—a condition that is becoming increasingly common (see Finney 1973; Howlett 1973; Howlett et al. 1976; Strathern 1982a; Westermark 1982; Grossman [in press])—people are especially vulnerable to the consequences of the conflicts between subsistence and commodity production, because increases in the acreage devoted to cash crops will be made at the expense of land for subsistence gardens. Under conditions of minimal technological inputs—as is characteristic in the Highlands—more land is usually required to support a household using proceeds from the sale of commodities than would be needed if relying on subsistence food crops (Mitchell 1976; Harris 1978). In effect then, commodity production can increase land scarcity, which exacerbates population pressure on resources. At the same time, these Highlanders will be pressed further by a predicament common to peasants throughout the Third World—the prices of the foods they buy will rise in the long run in relation to the prices they receive for the commodities they sell (see Nietschmann 1979; Dewey 1981).[3] Another potential problem with the encroachment of commodity production on subsistence endeavors is the *type* of gardens or crops that are sacrificed when land is lost to commercial activities. For example, in the Hagen area of the Western Highlands Province, people plant coffee on fertile land once reserved primarily for preparing mixed-vegetable gardens, which traditionally provided several important crops including nutritious greens. The people are now forced to rely more on the less diverse sweet potato gardens as well as on store-bought foods (Strathern 1982a). Such unfavorable dietary changes are frequently linked with expanding rural commercial activity (e.g. Dewey 1979; Fleuret and Fleuret 1980; Dewey 1981).

RURAL ECONOMIC DIFFERENTIATION

Economic differentiation can be manifested in a variety of realms. Regional and rural-urban differentiation occur as certain areas grow at the expense of others. Increasing inequalities between men and women at the local level and among competing classes at the national level are other forms. This study has been concerned with the impact of the state and the commercial economy on rural economic differentiation among households within a peasant community whose members still retain direct control over their major means of production, land.

Although diversity in the standard of living among Kapanaran households is not marked, quantitative indices of consumption levels by themselves do not provide adequate evidence of economic differentiation (Bernstein 1979: 67). Instead, this process

[3] The Papua New Guinea government has attempted to alleviate this problem somewhat by limiting the rate of imported inflation through adjustments in the Kina's exchange rate, though the policy's impact on rural purchasing power is uncertain (Berry 1978). In addition, the Coffee Industry Board collects an export coffee levy, which varies with the price of coffee, to help support prices when they are low, but it cannot halt substantial price declines, such as the one that occurred in the early 1980s.

must be examined in relation to changing social relations of production. In Kapanara, differentiation has occurred as cattle bosses have exerted an unprecedented degree of control over relatively large tracts of land, have appropriated a disproportionate share of the surplus from the enterprises, and have mobilized labor successfully without adequate reciprocation for several years (though there are eventual limits as to how much labor helpers will provide under such circumstances). In contrast, current big-men, by virtue of their status as patrilineage cluster leaders, could not exercise such a degree of control over land, labor, and surplus.

Economic differentiation in the village is in a relatively early stage, and certain contemporary socioeconomic patterns constrain growing inequalities. One is continuing investments in social relationships; some of the money earned from cattle and coffee is still spent in the system of reciprocal exchange either in the form of cash or goods. Differentiation is also contained as villagers allocate revenues from commodity production for household consumption. Commercial reinvestment and accumulation, crucial elements in furthering differentiation, do occur but are limited in nature. Villagers invest in trade stores, and several enterprises are quite profitable, but no one owns more than one store and people do not expand the size of their operation beyond the current small scale. An additional constraint on accumulation through local retailing is the lack of interest charges on customer debts, which are usually for small amounts in any case. Kapanarans also occasionally purchase a truck to transport fee-paying passengers, with the cattle bosses usually being the major investors. For example, in 1977 six men pooled money to purchase a vehicle; the three cattle bosses involved contributed K840 of the K1,000 collected. Some men in other villages in the Kainantu region who have purchased a truck (PMV) to transport passengers and engage in roadside coffee buying have earned substantial incomes and have reinvested their profits in other commercial endeavors, but the Kapanarans to date have not been successful financially in this activity because they lack the necessary technical and managerial skills (see also Finney 1973).[4] Furthermore, the employment of wage labor, which permits an increase in the scale of operations and the extraction of surplus value, is relatively infrequent in the village production process, being used occa-

[4] Nevertheless, the PMV operators can be successful "socially" in gaining prestige by participating extensively in *bisnis*.

sionally mainly by one cattle boss and a few men with comparatively large coffee holdings.

In contrast, in some other Highlands communities rural economic differentiation is much more advanced (see Finney 1973; Howlett *et al.* 1976; McKillop 1976a; Connell 1979; Gerritsen 1979; Good 1979; Fitzpatrick 1980; Good and Donaldson 1980). Several researchers have described affluent "big peasants" (Gerritsen 1979) who control many more resources and employ more wage labor (see Howlett 1980) than any Kapanaran. Such peasants have extensive commercial holdings, with interests in cattle projects, several hectares or more of coffee, transportation, restaurants, large market gardens, impressive trade stores, and other enterprises. Some have also managed to obtain large blocks of land either through tenure conversion (Ward 1981) or through government leases (McKillop 1976a), giving them secure individual title to the land and facilitating the procurement of bank loans.

Nevertheless, in analyzing rural economic differentiation, "[w]hat is of interest is not the stage which has been reached but the processes involved" (Raikes 1978: 290). In contrast to findings elsewhere (e.g. Finney 1973; Gerritsen 1979; Good 1979), I have argued that in the Kapanaran case the manipulation of traditional leader-follower relations has not been a major influence on differentiation. I have stressed the fundamental role of state policy. Furthermore, the significance of state policy and the potential for differentiation are affected by commercial market forces.[5]

In pursuing its goal of expanding the rural commercial economy, the government introduced smallholder cattle projects. The policies of government agents—the PNGDB and agricultural extension officers—in relation to the projects have facilitated differentiation. The availability of credit from the PNGDB enabled the rapid procurement of an asset far greater than the villagers were capable of obtaining by themselves in such a short period. Official stress on enclosing large areas within a barbed wire

[5] Debates concerning whether the state or the penetration of the commercial economy is more important in affecting rural economic differentiation (e.g. Williams 1976; Bernstein 1979; de Janvry 1981; Harriss 1982) are likely to be inconclusive because state policy intimately influences the functioning of the commercial economy and government action varies considerably from one country to the next. In addition, the processes of differentiation that are being examined—increasing male-female inequalities, regional variations, proletarianization, and household changes within peasant communities—can all be affected differently by market and state.

boundary fence and the Clan Land Usage Agreement assisted cattle bosses in gaining control over an unprecedented amount of land. Consequently, if more productive technologies or enterprises requiring access to consolidated holdings are introduced in the future, the bosses will be in a more favorable position than other villagers to adopt them.

Extension policy emphasized a "progressive farmer strategy" in which extension assistance was concentrated largely on dynamic village leaders and entrepreneurs (McKillop 1976b). Agricultural extension agents believed that progressive farmers would help diffuse information about commercial activities to other villagers and that the farmers' financial success would stimulate other envious community members to increase commodity production. An additional justification of the policy was to maximize the impact of scarce extension resources. Cattle bosses were among the prime beneficiaries of this policy; even though only a small minority of the country's population, they received more extension inputs than any other group during the 1970s (McKillop 1975). Servicing cattle projects and processing material related to PNGDB loans consumed much of the extension agents' time. The officers delivered cattle, provided veterinary services, helped organize and run the Kainantu Bulumakau Association, often directed project members to make improvements, and helped in marketing cattle—all inputs that assisted the bosses. This concentration of extension effort on cattlemen was widespread in the Highlands (Howlett *et al.* 1976: 247-250; McKillop 1976a, b; Grossman, 1980).

The exclusiveness of the links between the cattle bosses and state agencies also conferred an air of legitimacy on the cattle bosses, creating the impression of official government support. Consequently, the bosses' position within the village was strengthened. At the same time, the differentiation process was not simply imposed from outside. Project leaders themselves were particularly adept at manipulating their privileged access to the government bureaucracy, selectively interpreting and disseminating information from the outside world to gain control over land, labor, and wealth (Grossman 1983).

The role of the state in stimulating economic differentiation within rural communities is common in the Third World. Many colonial governments in Africa, for example, adopted a progressive farmer strategy (Williams 1976; Raikes 1978; Heyer *et al.* 1981) with the goal of establishing an economically and politically stable group of commodity producers, an obvious benefit to the state. The provision of subsidized credit and technological inputs such as fertilizers, pesticides, irrigation, and agricultural machinery aided only a minority of the rural population and discriminated against small landowners (see Henderson 1972 for a similar situation in Belize). State policy can speed differentiation not only by assisting rich peasants but also by undermining poor ones through such means as excessive taxation (Shenton and Watts 1979), the reduction of prices peasants receive for their crops (Bates 1981), or land expropriation for establishing large-scale capitalist farming enterprises (Taussig 1978).

Government policy concerning land ownership is especially influential because the introduction of freehold tenure making land a marketable commodity enables rich peasants to consolidate their holdings and rapidly expand the scale of their agricultural operations as some poorer peasants sell part or all of their land (Mafeje and Richards 1973). As the case from Kapanara demonstrates, differentiation can certainly occur in the absence of legally marketable land, but its extent will be limited. Nevertheless, the process can still proceed much further than it has in Kapanara or elsewhere in the Highlands. For example, in Ghana, where land also is not a legally marketable commodity, poor peasants commonly have to mortgage their land or crops at usurious rates, which is the major force contributing to rural differentiation there (Howard 1980).

State policy and cattle bosses' manipulation of links with government agencies have resulted in an initial boost to differentiation. The commercial economy has created additional forces abetting this process. Although the expansion of commodity production, by itself, will not inevitably lead to the dissolution of the peasantry into capitalist farmers and wage laborers (Shanin 1979), it nevertheless can undermine previously established patterns of social interaction that contain the rate of differentiation.

The spread of the commercial market is often linked directly to the differentiation process through such means as worsening terms of trade for rural producers, the vulnerability of poor households to market uncertainties, the resulting classic cycle of indebtedness and impoverishment, growing wage labor, and the eventual consolidation of land holdings. Another important consequence of expanding rural commercial activity that is particularly relevant to this study is increasing individualism, which refers to the gradual decrease both in the range of people granted certain customary rights and in the number of community members with whom certain traditional ties are maintained. In essence, individuals or groups attempt to reduce the customary access and rights that others have in their resources and to retain maximum economic advantage for themselves. These changes are manifested in a variety of social realms such as land tenure and reciprocal patterns of food and labor exchange. The impact of increasing individualism on different segments of a community varies. It benefits mainly the wealthy, because by weakening customary communal ties that previously limited the rate of capital accumulation by particular households, it facilitates rural economic differentiation. At the same time, it is most detrimental to poor households because individualism attenuates the reciprocal networks that provide the poor with social security in times of need. In the Kapanaran case, it has been significant largely in terms of its impact on the wealthy, because the impoverished households that characterize many other Third World peasant communities are absent in the village.

Certainly, commodity production is not solely responsible for individualism in the Highlands or other Melanesian communities because individualism has always been an essential part of the social fabric. Langness (1968) has called attention to the significance of individualism as opposed to corporate group actions in the traditional political process in the Highlands, and Hogbin (1939: 169), in response to claims that European contact and a money economy made Solomon Islanders into individualists, proclaimed that "these natives have always been individualists." Thus, market forces accentuate an aspect of traditional Melanesian life.

At the same time, to stress exclusively the importance of individualism while ignoring the communal aspects and reciprocity characteristic of Highlands communities is one-sided and misleading. The orientation toward groups and mutual assistance was adaptive in the context of intergroup warfare and periodic food shortages, which existed before pacification. Individuals needed others to survive. The traditional system of reciprocity usually ensured that the effects of shortages were distributed widely in the community to reduce the burden on particular households. Today, although reciprocity and group endeavors are still evident, many pressures people experienced previously are absent. Some Kapanarans now view traditional obligations of reciprocity as a hindrance to their success in *bisnis* and capital accumulation, as is evidenced in the cattle bosses' failure to compensate their followers adequately for their assistance.

The trend towards more individual forms of land tenure in areas where commodity production is extensive is a common pattern in Papua New Guinea (Harding 1972; Ward 1981). Mitchell's (1976: 81) findings concerning the impact of commodity production on the Nagovisi of the North Solomons Province are also applicable to many other areas in the country:

> Land, never an important commodity in the absolute sense, has become a scarce resource in some groups and is now perceived by all—land-rich and land-poor—as a valuable entity in the abstract.

Similarly, Kapanaran attitudes toward land values are changing. Before the advent of commodity production, differential control of or access to land in Kapanara was not important in determining success in economic or political activities because of traditional constraints on production and the abundance of land. Thus, members of one patrilineage cluster did not to any great extent attempt to restrict others in the village from establishing proprietary rights on the cluster's land. However, villagers now perceive land as being a scarce resource because so much has been enclosed by the cattle projects. Other men also wanted to start cattle projects but did not have enough land. People are thus aware of the increasing economic value of land and want to restrict others' access to their land because multiple claims to a tract dilute its revenue potential. Consequently, some villagers now assert that one should obtain proprietary rights only in the territory of one's own patrilineage cluster, and in several cases members of one cluster prevented those of another from establishing such rights. Control over land has many potential economic advantages, enabling those with access to larger holdings to expand commercial output more readily and to earn money from selling land or timber to the government. One villager neatly summarized the changing perspective: "Before we were not clear about these matters. Now we understand. Everything has a cash value."

Growing individualism in relation to land is linked most directly to economic differentiation in Kapanara through the cattle bosses' attempts to gain exclusive control over the enclosed areas, to the detriment of other project members. They now claim that the enclosed land is theirs and will be inherited by their sons, even though the land was initially part of a larger group's estate.

The manifestation of such individualism in what were once communal or group endeavors is widespread in many types of commercial activities in the Highlands (Finney 1973), a change that clearly enhances the economic position of selected villagers.

Patterns of reciprocity and cooperation have also been affected by commodity production. The extent of mutual assistance and the degree to which individuals have access to resources other than those controlled by their own household are lower in commodity production than in subsistence activities. To examine these changes, I use data from the time-allocation study, comparing patterns for married adults in subsistence food gardening and coffee production. For each observation, I noted whether individuals were working with members from other households to indicate the degree of mutual assistance. Mutual assistance occurred in 13 percent of the observations of subsistence gardening but in only 5 percent of those involving coffee production. To examine the degree to which individuals had access to resources other than those controlled by their own household, I recorded whether people were using other households' resources (either for their own benefit or for that of a household other than the one controlling the resource). In 15 percent of the observations of subsistence gardening, individuals used resources other than those controlled by their household; they did so in only 3 percent of the cases involving coffee production. These indicators of patterns of assistance and resource use reveal a greater restriction of reciprocity, cooperation, and sharing in commodity production.

Such differences are reflected in the ideological separation of subsistence and *bisnis*, with each domain having its own system of morality. When I asked one trade store owner why he did not share food from his store as he did food from his subsistence gardens, he declared in *Tok Pisin*, *"Bisnis, em i narapela samting"* (*"Bisnis* is a different matter"), indicating that appropriate behavior in income-producing activities is distinguishable from that in traditional subsistence endeavors. Whereas in subsistence activities, instances of a breakdown in expected reciprocity are rare, in commercial activities they are legion. Complaints about not being adequately rewarded for investing in trade stores or cattle projects are common not only in Kapanara but in many Highlands communities. Failure to reciprocate, an indication of increasing individualism, enables some villagers to retain more revenues for themselves.

Certainly, increasing individualism associated with the spread of the commercial economy is not unique to the Highlands. As one Tahitian observed, "Kinsmen no longer look after kinsmen, they care only for money" (Finney 1965: 3), and the Ashanti of Ghana say, "Cocoa kills the family" (Boyon 1957, cited in Stavenhagen 1975: 89). Declines in reciprocity and group endeavors and growing individualism in land tenure, which facilitate differentiation, occur throughout the Third World (see Finney 1965; Nietschmann 1973; Stavenhagen 1975; Lingenfelter 1977; UNESCO/UNFPA 1977; Gudeman 1978; Klein 1980). In some cases, these changes are in response to increasing impoverishment, as those less well off can no longer fulfill traditional obligations (see Shenton and Lennihan 1981). In contrast, in Kapanara increasing individualism reflects the powerful lure of money and the *bisnis* ethos.

PERSISTENCE OF THE PEASANTRY

Although Kapanara's incorporation into the commercial economy is very recent by world standards, its unfolding story can still illuminate the fundamental issue of the persistence of the peasantry. In focusing on this topic, I emphasize the forces affecting rural involvement in commodity production.[6] Although the intensification of commodity production would not inevitably create a group of capitalist farmers and landless (or nearly landless) wage laborers,[7] it would nevertheless lead to major transforma-

---

[6] Much of the literature on the persistence of the peasantry deals with the benefits that capitalists derive from the survival of peasant communities and with the factors enabling peasant households to compete with large-scale farming enterprises. For a sampling of this literature, see Meillassoux (1972); Cliffe (1976); Williams (1976); Stavenhagen (1978); Taussig (1978); Vergopoulos (1978); Shenton and Lennihan (1981); and Harriss (1982). In the Papua New Guinea context, see Fitzpatrick (1980). Researchers contend that the maintenance of peasant communities is beneficial to capitalist interests because peasants supply inexpensive produce and labor for the commercial economy. Peasant communities subsidize capitalist interests because village subsistence production and reciprocal networks provide part of the costs of rearing labor as well as social security for its members. Thus, capitalists can give them low prices for their products and labor because merchants and employers do not have to pay for the entire costs of raising peasant families and maintaining them after retirement, as they would for full-time laborers. Furthermore, the persistence of peasant communities maintains ethnic boundaries and consequently counters class consolidation that would challenge capitalist interests (*ibid.*).

[7] Moneylenders may prefer not to obtain control of land belonging to debtors who fail to repay loans fully, instead keeping them in a perpetual state of in-

tions in production patterns. As market involvement grew in the Highlands, subsistence production would decline, economic differentiation would widen, and most likely some segments of rural communities would become impoverished and be forced into more wage labor and, possibly, eventually financial indebtedness.

The likelihood of such a situation in Kapanara and the rest of the Highlands in the immediate future is lessened considerably by a basic fact of life in the Third World—boom and bust cycles related to fluctuating commodity prices[8] and to varying government inputs supporting peasant commercial agriculture. Had commercial involvement in Kapanara continued to increase as it had in the mid-1970s, such major changes would be possible, but the economic bust of the early 1980s helped constrain the rate of change in productive activities. Wolf's (1955) remarks concerning Latin American peasantries indicate that the conditions Kapanarans faced in a relatively short period are characteristic of the long-term experiences of many peasants elsewhere:

> It would be a mistake, moreover, to visualize the development of the world market in terms of continuous and even expansion, and to suppose therefore that the line of development of particular peasant communities always leads from lesser involvement in the market to more involvement. . . . During the second part of the nineteenth century and the beginning of the twentieth, many Latin American countries were repeatedly caught up in speculative booms of cash crop production for foreign markets, often with disastrous results in the case of market failure. Entire communities might find their market gone overnight, and revert to the production of subsistence crops for their own use.

To understand Kapanara's reaction to such fluctuations requires considering the three variables affecting involvement in com-

---

debtedness that obligates the peasant to sell his produce to and buy goods from the moneylender at unfavorable rates (Elwert and Wong 1980: 516-517). Also, wealthy farmers and moneylenders may decide not to reinvest profits in agricultural production when the rates of return are higher in nonfarm ventures, thus limiting accumulation and differentiation within the agrarian sector (Bernstein 1979; Harriss 1982). Consequently, poor peasants may not be forced off their land as commodity production increases.

[8] Fluctuating "prices" are influential in a relative, not absolute sense. Thus, if commodity prices increase, but the cost of purchased items rises more rapidly, little incentive may be provided to produce more cash crops and a decline in commodity production may occur (e.g. Beckman 1981: 157).

modity production: the degree of risk posed by expanding commercial activities, the nature of political-economic forces, and the extent of local autonomy from other classes, the state, and the market.

With their production systems being mostly resource-based and having a low level of technology, peasants often face serious local environmental problems, making commitments to commodity production risky. Local environmental problems (and therefore the risks) are often exacerbated by political-economic conditions, as Scott (1976: 1) reveals in his portrayal of the plight of colonial peasants in Southeast Asia: "Here too, lilliputian plots, traditional techniques, the vagaries of weather and the tribute in cash, labor, and kind exacted by the state brought the specter of hunger and dearth, and occasionally famine, to the gates of every village." In contrast, the condition of "subsistence affluence" that has characterized the Kapanaran production system since pacification (but is now being undermined by the commercial economy) reflects comparatively fewer local environmental constraints. Although a distinct dry season occurs, villagers can surmount the problem by planting a variety of crop combinations in the different environmental zones throughout the year. Increased commodity production has made the villagers more susceptible to local environmental problems, but the Kapanarans do not view local environmental conditions as a major obstacle to expanding commercial activity.

The nature of their political-economic environment is partly conducive to intensifying cash-earning activities. Kapanarans are not hindered by the exactions of a landlord or moneylending class or by forced land expropriation. The state has facilitated rural output through agricultural extension activity and the granting of PNGDB loans. Furthermore, villagers generally have a favorable view of agricultural extension agents, believing that the cash-earning activities they introduce will help bring "progress."[9] In contrast, in other areas such as in Africa, extension agents have been associated with such extremely unpopular programs as destocking, cattle dipping, erosion control, and mandatory grain reserves, making peasants there hesitant to accept their advice to expand commodity production or adopt innovations.

[9] Nevertheless, Kapanarans have not viewed all government actions concerning agriculture positively. For example, past attempts to limit burning grasslands and to ban cultivating at the forest fringe were unpopular in the village.

However, two crucial elements in Kapanara's external, political-economic environment—the price of coffee and the extent of agricultural extension inputs assisting cattle projects—have varied over time. In the mid-1970s, they stimulated increased commodity production, which, in turn, had extensive ramifications at the local level—the strengthening of the *bisnis* ethos, disruption and decline in subsistence production, environmental deterioration, growing dependence on imported foods, increased individualism, and widening economic differentiation—consequences that then led to even greater commitments to the commercial sector. But the transforming power of these processes was muted in the early 1980s, as the drop in coffee prices and the decline in agricultural extension assistance resulted in Kapanarans decreasing their involvement in the cash economy while expanding subsistence production. Although such fluctuations in external, political-economic conditions are commonplace, the impact that they have on peasant communities will vary depending to a large extent on the degree of local autonomy.

Kapanarans have considerable autonomy from other classes, the state, and the market. Cash requirements for taxes, school fees, and purchased foods are still low. Land for subsistence production remains adequate. The only major debt in the village is to the PNGDB for the cattle projects, but the Bank holds a mortgage on the stock and improvements (such as fencing material) only and the interest rate is commercial, not usurious. Other Highlanders who have defaulted on such loans have not lost control of their land or other nonproject productive resources. Inflation in ceremonial expenses such as bridewealth (approximately K200 in 1977) and *kandere* exchanges has created the need for more money, but the impact on individuals is minimized somewhat because many people often pool their cash in making the payments. Consequently, although certain forces do require a commitment to the commercial sector, they are not extensive. Thus, to a substantial degree, Kapanarans can increase or decrease commodity production as conditions warrant.

The level of local autonomy affects not only the way commodity production changes in response to fluctuations in the external, political-economic environment, but also the potential for economic differentiation within rural communities. Where the degree of local autonomy is low, peasants confronted with declining crop prices (or terms of trade) can be caught in the simple reproduction squeeze—compelled to intensify inputs into unrewarding commodity production and wage labor while faced with declines in both subsistence food output and cash income—possibly forcing some poor peasants into a cycle of indebtedness and even eventual loss of land. The increasing impoverishment of these poor peasants benefits rich peasants, merchants, and moneylenders, thus facilitating economic differentiation. In contrast, such impoverishment that abets differentiation during economic downturns did not occur in Kapanara because the villagers, having a high degree of local autonomy, were able to reduce inputs into the unrewarding commercial economy during the bust of the early 1980s for several reasons. The severity of the food shortage occurring during the bust was ameliorated somewhat by the continuing viability of reciprocal exchange networks; villagers thus were not forced to expand commercial activities to make up for reduced food production because they could rely to some extent on gifts of food from other Kapanarans who were better off and from relatives in nearby villages. People also retained adequate access to land and other productive assets necessary to supply household consumption needs. Furthermore, because usurious debt was absent and cash requirements were not burdensome, no one was forced into the cycle of increasing poverty.

Kapanara's autonomy is not based exclusively on the ability to reduce the amount of time allocated to commodity production. The nature of the productive enterprises themselves are important. Not only can the villagers restrict inputs into cattle raising, but they can remove the projects themselves to establish subsistence gardens, as has occurred in a few densely populated Highlands areas. Although the Kapanaran bosses would undoubtedly attempt to prevent such an outcome, it remains an option. The planting of coffee, a perennial tree crop, more effectively removes land from subsistence cultivation for many years. People would be hesitant to uproot their trees to plant food gardens because coffee, which only starts to bear a crop three or four years after planting, represents a considerable investment of time and land. Nevertheless, they could remove the trees if conditions required, as have some coffee-growing peasants in response to severe land shortage (in Maragoli area of Kenya—Cowen 1981: 64) and to the decline of markets (during World War II in Brazil—Payer 1975: 159).

As the case study of Kapanara reveals, political-economic forces are the major influence on peasant autonomy, but other factors such as the nature of the commodity produced and changes in

ceremonial expenses also affect the reversibility of the commitment to commercial activities. In addition, although resource
depletion (in this case, soil erosion) did not significantly affect
Kapanara's autonomy, it has played a more crucial role elsewhere.
These other factors have generally received less attention in the
development literature than political-economic forces, but nevertheless can have a fundamental impact on local autonomy.

Case studies from Panama (Gudeman 1978) and northeastern
Brazil (Gross and Underwood 1971) further illustrate the manner
in which the nature of the cash crop influences reversibility in
rural communities. In contrast to the Kapanaran situation, commodity production in these two cases had an almost irreversible
impact, negatively affecting subsistence endeavors.

In rural Panama, population growth and the planting of a drought-
resistant pasture grass that hindered forest regeneration contributed to a decline in the forested area per capita. Because peasants
cultivated gardens in the forest, their subsistence base was also
diminishing. Pressed by decreasing natural resources and lured
by the desire for money, which villagers hoped would provide
"the long-awaited opportunity for participating in 'civilization' "
(Gudeman 1978: 131), they adopted sugar cane as a cash crop.
Planted in food gardens, cane was easily assimilated into the traditional agricultural cycle, and initially it appeared a solution to
their problems. However, sugar cane subsequently added to the
problem of resource depletion. Unless plowed, sugar cane will
continue to grow after harvesting, thus preventing forest regeneration and reducing the amount of land available for subsistence
even further. Pressed by a continually diminishing subsistence
base, some peasants began using fertilizers and plows to increase
cane production. However, these "modern" techniques had to be
purchased with credit from the cane-processing mills, and repaying the credit required the planting of more cane, thus exacerbating the decline in natural resources. The only way to stay even
was to plant still more cane—to the detriment of subsistence
production.

In northeastern Brazil, peasants enthusiastically adopted sisal
as a cash crop in the drought-prone area because it is drought-
resistant, the government encouraged its planting, and at the time
its market price was high. Some planted most or all of their land
with the cash crop, causing a widespread abandonment of subsistence agriculture and forcing them into wage labor while waiting for the sisal to mature. When their crop was ready to harvest

four years after the initial planting boom, the price of sisal had dropped substantially. Further exacerbating their problem, owners of sisal-processing machines discriminated against such small-holders because of the lower profitability in servicing small areas. The commitment to sisal was nearly irreversible because once planted, the crop is very difficult to eradicate. The men, unable to return to subsistence production, had to work at such strenuous jobs that they had to deprive their children of food to obtain enough energy to continue working (Gross and Underwood 1971).

Inflation in ceremonial expenses, a widespread phenomenon accompanying the spread of commodity production, has increased the need for cash in many peasant communities, for without sufficient access to money people cannot participate fully in culturally meaningful and socially necessary activities such as marriage. Brookfield (1968a: 100) argued that the growing need for money in inflating, competitive ceremonial exchanges was a major factor stimulating the Simbu of the Highlands to expand coffee production in the 1960s. Among the Hausa of northern Nigeria, the costs of naming ceremonies, gift exchanges, and bridewealth have risen substantially (Watts 1979). Their burdensome marriage expenses frequently necessitate large-scale borrowing and are a principal reason for land sales and pledging. By contributing to indebtedness, they help bind peasants to the cash economy and, along with exploitative, political-economic forces, reduce local autonomy.

As villagers increase commercial output to satisfy growing cash requirements, resource depletion may result, further tying them to the commercial economy to the detriment of subsistence production. The Miskito Indians of Nicaragua once relied on turtle mainly as a source of food, catching them only during certain seasons (Nietschmann 1979). However, with the arrival of turtle-processing companies, the demand for turtle as a commodity rose sharply. To facilitate harvesting, the companies extended credit and supplies to the Miskito to enable them to catch turtle the entire year instead of seasonally. As people began to spend more time on procuring turtles and on other cash-earning activities, subsistence agriculture declined, and thus a cash flow had to be maintained to purchase food and supplies to make up for the reduced subsistence output. At the same time, the intensified pressure caused the turtle population to decline, but because villagers had to buy more food and repay their loans, they subsequently had to spend even more time searching for the increasingly scarce turtles. The Miskito thus became locked into a positive feedback system—spiraling declines in both the turtle population and subsistence output, a situation exacerbated by worsening terms of trade. Pressed by resource depletion, disruption of subsistence endeavors, and growing cash needs, the Miskito eventually were forced to engage in increased wage labor migration. In essence, resource depletion contributed to a decline in local autonomy, making it difficult to reverse the commitment to the commercial economy.

Although decreasing local autonomy engenders certain serious problems for peasants, some researchers nevertheless view peasant autonomy as detrimental to national welfare. Indeed, Hyden (1980) has gone so far as to assert that enduring peasant autonomy—in contrast to exploitative, political-economic forces—is the major cause of underdevelopment in Africa.[10] He contends that peasants in Africa, in contrast to those in Latin America and Asia, maintain a considerable degree of independence from other social classes, the state, and the commercial market because they retain control over their means of production, their subsistence endeavors remain viable, and their system of social relations provides security. In addition, although they participate in the market, they are minimally dependent upon it. He argues that African peasants value their autonomy and are too concerned with time-consuming subsistence and community social endeavors, making them reluctant to make changes in their productive activities—such as expanding commodity production or adopting mechanized technologies that would link them more closely with the market and state. Consequently, because peasants with autonomy have an "exit option," they can resist government demands to produce the increased farm output required to feed growing rural nonfarm and urban populations and to achieve national self-reliance. Thus, Hyden views the loss of peasant autonomy as a crucial step toward development, a perspective summarized aptly by Williams (1976: 148):

Peasant households control and manage their means of production, thus allowing them a measure of autonomy vis-à-vis

[10] This viewpoint is clearly in opposition to my own perspective. In his analysis, Hyden fails to give due consideration to the role of political-economic forces in undermining peasant welfare. Problems associated with colonialism, usurious debt, and the simple reproduction squeeze—which limit peasant autonomy—have contributed to underdevelopment in many parts of Africa.

other classes. This autonomy must be broken down if peasant production is to be adapted to the requirements of urban, industrial capital formation and state development planning. Peasants must be made dependent on external markets and power-holders for access to the resources which come to be necessary to their way of life, or they must be coerced into organizing production to meet external requirements.

In contrast, I assert that the maintenance of local autonomy is essential for the welfare of peasant communities. Autonomy implies, among other things, the existence of a viable subsistence system, and it is subsistence production that provides the crucial buffer insulating peasants from inevitable national and international economic cycles. The fluctuation of cash crop prices, particularly for export commodities, is a widespread problem for Third World countries, and even when governments intervene to help stabilize prices, peasants' returns are generally low, as governments often heavily tax agricultural commodities to support development in other sectors (Lele 1981). By retaining autonomy, peasants can shift back into more subsistence production as commodity production becomes less rewarding. If the subsistence system is adequate, the reversal enables peasants to provide themselves with sufficient food during economic downturns. Although such an outcome may seem very modest and limited compared to the enticing promises of governments and planners that increasing affluence will follow the expansion of rural commercial activity, it is nevertheless a substantial accomplishment when compared with the reality of growing hunger in much of the Third World.

As the Kapanaran case demonstrates, the shift back into subsistence production is not necessarily a smooth transition. For example, reduced coffee inputs contributed to a cash shortage, and decreased time spent on cattle raising led to increased garden damage.

But if the transitional period is not problem free, the alternative can be much worse. Peasants with a low degree of autonomy are forced to intensify unrewarding commodity production or wage labor when producer prices (or terms of trade) decline, often leading to material impoverishment—the simple reproduction squeeze. However, the problem with the simple reproduction squeeze is not only that falling prices necessitate increased commercial activity to the detriment of labor and land inputs into subsistence endeavors, but that expanded commodity production itself can lead to a series of interlocking social, economic, and ecological processes that function as positive feedback cycles continually undermining subsistence production. In essence, being sucked into the vortex of the commercial economy can seriously erode community welfare. The promised land that commodity production is supposed to deliver often turns out to be an illusion.

Ecological Methodology

VEGETATION TRANSECTS

In examining vegetation change, the basic unit of analysis was a meter-square frame, which was moved along the transect line. Within the square, the percentage of the area covered by the leaves of each plant species was recorded in classes from:
  (1) 25 percent or less
  (2) 26 to 50 percent
  (3) 51 to 75 percent
  (4) 76 to 100 percent
The coverage of most species falls within the first class. The frequency of a species is the number of times it was present (each time the frame was moved, the presence or absence of a species was noted) for each 100 square meters examined. I use the term "occurrence" synonymously with frequency. These data are useful for examining changes in individual species. For changes in the total vegetative cover, the data on the percentage leaf coverage beneath the transect line are more useful. It is common to visually overestimate leaf coverage, but because the data are comparative and the same bias is present in each case, no problem arises.

In the overgrazed project, seven transects were made consisting of between 50 and 100 squares for a total coverage of 450 square meters. In the areas that were ungrazed or very slightly grazed, six transects were made of 25 squares each for a total of 150 square meters. The areas analyzed have not been gardened for at least 20 years. The greater percentage coverage in the overgrazed project was necessary because of the greater variability in vegetation composition. Quantitative data were supplemented with casual observations.

BULK DENSITY

At the site to be sampled, the vegetation was clipped to ground level. Then the area was thoroughly saturated by lightly sprinkling water on the soil surface. The site was covered with plastic sheeting and cut grass laid on top to prevent any evaporation from the soil. Because clays expand to different volumes depending on the moisture content of the soil, all sites had to be saturated to ensure comparability. I returned to the sites 48 hours after saturation when field capacity was assumed to have been reached. Using a soil corer similar to the one described by McIntyre (1974), I obtained soil cores 4.95 centimeters in height and 8.56 centimeters in diameter by hand-driving the head of the machine using a rubber mallet. The soil was trimmed to conform to the size of the core. Four cores were usually taken at each site, but occasionally one of the four had to be rejected. The soil was later heated in an oven for 30 to 36 hours at approximately 110°C. Dried samples were weighed on an electronic balance.

Soil Erosion

Boards 15 centimeters wide were calibrated and placed upright in the soil with the face of the board perpendicular to the slope to trap sediment being washed or splashed downslope. Boards were placed at various angles and lengths from the top of the slope (see Table A.1) to obtain a representative sample of the areas. It was impossible to select sites exactly equivalent in slope angle and length to the top of the hill in the two zones because of variations in slope form, trails, localized disturbance by villagers digging for edible insect larvae, and rooting by pigs. The average length from the top of the slope to the erosion boards was slightly greater in the overgrazed area (68.26 meters) compared with the relatively ungrazed area (61.75 meters), but this difference is exaggerated because of the existence of cattle trails that are oriented across the slope in a slight incline. The erosion boards collected sediment carried by sheetwash down the hill, but in the overgrazed project sediment could not be transported the entire length of the slope because the cattle trails provided alternative paths for the sediment to travel. The average slope angle was slightly greater in the relatively ungrazed zone (23.66°) compared with the overgrazed area (22.82°). The distribution of the slopes and lengths to the top of the hill are roughly comparable (see Tables A.2 and A.3). The first six boards in each area were set in place on April 9, 1977, and the rest of the boards on April 19, 1977.

The level to which sediment accumulated at each site on different occasions is indicated in Table A.4. The data-collecting schedule was irregular because of the need to take readings after

TABLE A.1

Site Characteristics of Erosion Boards

Overgrazed Area			Relatively Ungrazed Area		
Board	Distance to top of slope (m)	Slope (degrees)	Board	Distance to top of slope (m)	Slope (degrees)
1	55	21	A	50	25.5
2	61	25	B	91.5	18.2
3	62	20	C	84	17.3
4	43	17.5	D	32	24
5	82	17	E	37	22
6	74	28	F	47	22.1
7	33.5	18	G	90	18.4
8	43.5	21	H	38	32
9	82	17	I	53	18.5
10	45	28	J	66	23.2
11	35	30	K	38	32
12	82	17	L	79	22.9
13	50	23	M	30	21
14	40	31	N	61	20.9
15	121.50	23.3	O	79	22.9
16	45	30	P	84	26.5
17	121.50	23.3	Q	84	26.5
18	121.50	23.3	R	43	23
19	103	27.3	S	70	25.8
20	42	21	T	70	25.8
21	71	23.9	U	66	27.8
22	121.50	23.3	V	66	24.5
23	35	16			
Mean	68.26	22.82	Mean	61.75	23.66
Range	33.5-121.5	16-31	Range	30-91.5	17.3-32

periodic heavy rains and because of short absences from the village. After each reading, the accumulated soil was cleared. Measurements that are positive in value have been added to yield the total level of accumulation. Negative results indicate that the level of the soil was below the original soil level. Negative readings have not been used to compute the totals because the loss of soil must have been the result of local disturbance created by the erosion board interacting with heavy runoff.

Several factors besides the existence of the cattle trails intro-

TABLE A.2

Distribution of Length to Top of Slope
for Erosion Boards

Length (m)	Overgrazed Area	Relatively Ungrazed Area
30	4	5
41-50	6	4
51-60	1	0
61-70	2	6
71-80	2	2
81-90	3	4
91-100	0	1
101+	5	0

TABLE A.3

Distribution of Slope Angle for Erosion Boards

Slope Angle (°)	Overgrazed Area	Relatively Ungrazed Area
16	1	0
17	4	1
18	1	3
19	0	0
20	1	1
21	3	1
22	0	4
23	6	2
24	0	2
25	1	3
26	0	2
27	1	1
28	2	0
29	0	0
30	2	0
31	1	0
32	0	2

TABLE A.4

Height (cm) to Which Soil Accumulated Against Erosion Boards
(Overgrazed Area)

| | | | | Date of Reading | | | | |
Board	19 April	8 May	28 May	22 June	4 August	27 September	6 December	Total
1	1.0	1.0	3.0	0	2.3	0.5	2.0	9.8
2	2.0	0	0	0.5	0	1.5	4.0	8.0
3	2.5	KD	2.0	1.0	1.0	7.0	2.5	16.0
4	KD	1.0	1.0	0.9	0.5	KD	2.5	5.9
5	0.5	0.5	2.5	0.5	0.3	0	0.5	4.8
6	KD	0.5	1.5	−1.0	KD	7.0	7.0	16.0
7	0	0	0.3	0	0.2	CD	PR	0.5
8		1.0	0	1.5	0.3	0.2	PR	3.0
9		1.0	0	1.0	0	0	1.0	3.0
10		−0.3	0	−0.2	CD	−0.5	PR	0
11		−0.3	0.3	−0.1	0	0.5	PR	0.8
12		3.0	0	1.8	0.3	1.0	2.5	8.6
13		1.9	0.1	1.9	0	0.3	0	4.2
14		1.5	1.7	2.0	1.0	0	1.0	7.2
15		3.0	1.0	2.9	−2.0	2.0	2.0	10.9
16		2.0	0	2.9	0.5	0	0.9	6.3
17		0	0	1.0	3.0	1.0	0.5	5.5
18		0	1.0	0	KD	0	KD	1.0
19		1.3	0	0	0.8	2.0	2.0	6.1
20		0	−0.5	0	PR	1.0	0.5	1.5
21		4.0	3.5	5.4	4.0	1.0	5.0	22.9
22		3.0	1.8	2.7	0	0.3	4.0	11.8
23		0	1.0	0	0.9	0	0	1.9
Total	6.0	24.7	20.7	26.0	15.1	25.3	37.9	155.7

KD = knocked down by cattle   CD = cattle disturbed site   PR = pig rooted nearby

duced a bias lowering the readings for the overgrazed area. Sometimes, cattle trampled next to the boards or pigs rooted near the boards causing abnormally large accumulations. Occasionally, cattle knocked over the boards. In all such cases, the readings were disregarded. In cases of pigs rooting nearby, the erosion boards were moved away from the disturbed site and placed at a comparable site nearby to prevent further abnormal readings. Because of the magnitude of the difference between the level of accumulation in the two zones, no attempt is made to compensate

TABLE A.4 (*cont.*)

Height (cm) to Which Soil Accumulated Against Erosion Boards

(Relatively Ungrazed Area)

				Date of Reading				
Board	19 April	8 May	28 May	22 June	4 August	27 September	6 December	Total
A	0	0	0	0	0	0	0	0
B	0.9	0.1	0.8	0.1	0	0	0	1.9
C	0	0	−0.1	0	0	0	0	0
D	0	0	−0.3	0	0	0	0	0
E	0	0	0.5	0.5	0	0	0	1.0
F	0.1	0	0.1	0	0	0	0	0.2
G	0	−0.1	0	0.5	0.5	0.1	0.3	1.4
H		0	−0.1	0.1	0	0	0	0.1
I		0	−0.1	0	2.5	1.0	0	3.5
J		0	0	0	0	0	0	0
K		0	0.5	0.4	0	0	0	0.9
L		0	0	0.1	0	0	0	0.1
M		0	0.3	0.2	0	0	0	0.5
N		0	0	0.3	0	0	0	0.3
O		0	0	0	0	0	0	0
P		0	0	0	0	0	0	0
Q		0	0	0	0	0	0	0
R		0	0	0	0	0.1	0.1	0.2
S		0	0	0	0	0	0	0
T		0	0	0	0	0	0	0
U		0	0	0	0	0.1	0	0.1
V		0	0	0	0	0	0	0
Total	1.0	0.1	2.2	2.2	3.0	1.3	0.4	10.2

for the lost readings in the overgrazed zone. The bias is somewhat balanced because 23 boards were set in the overgrazed area, whereas only 22 were placed in the other area. In computing the total level of accumulation, 123 cases are added from the overgrazed area compared with 134 cases for the relatively ungrazed zone, thus accentuating the level of accumulation in the relatively ungrazed area and underestimating the magnitude of the difference between the two zones.

Selection of Sample of Households

A small sample had to be chosen because the collection of certain types of data, particularly those concerning time allocation and the area under cultivation, is extremely time-consuming. The sample comprised thirteen households; ten were headed by married men and the other three by widowed, elderly adults (two males, one female). I initially wanted to compare the households of cattle bosses with those of other villagers as well as to obtain information reflecting general patterns in the village economy. Half the married households in the sample were headed by cattle bosses to facilitate the comparison of cattle boss with non-cattle boss households. Although this gave my sample a ratio of 1:1 cattle boss to non-cattle boss married households, as opposed to 1:6 for the village population as a whole, the validity of my sample in terms of reflecting economic patterns of the village is still considerable as two of the five cattle bosses rarely worked on their cattle project.

The ten households headed by married men were chosen primarily on the basis of the ratio of the number of consumers to workers in the household. Chayanov (1966) suggested that the higher a household's consumer-worker ratio, the greater the intensity of its workers' inputs into production. Although other factors such as the desire for prestige (Sahlins 1972) and a commitment to cash-earning enterprises also influence a household's inputs into productive activities, the consumer-worker ratio is nevertheless a useful criterion by which to stratify the Kapanaran population. Because an examination of subsistence production was an integral part of my study, it was necessary to select households with consumer-worker ratios reflecting the distribution of ratios in the village. The consumer and worker units employed to stratify the population (see Tables B.1 and B.2) were based on observations in the field and on reports on other Highlands societies (e.g. Rappaport 1968: 75; Waddell 1972: 221; Boyd 1975: 149-51). In addition, pigs are consumers because they are fed garden produce. In calculating the ratio, small pigs are considered as

TABLE B.1

Consumer Units

	Age					
Sex	0-2	3-4	5-9	10-14	15-54	55+
Male	0.3	0.4	0.5	0.7	1.0	0.6
Female	0.3	0.4	0.5	0.7	0.9	0.6

TABLE B.2

Worker Units

	Children*			Un-married Adult 15+	Married Active	50-59	Semi-Productive Elderly	Non-Productive Elderly
Sex	0-6	7-9	10-14					
Male	0	0.1	0.3	0.5	1.0	0.7	0.5	0
Female	0	0.2	0.4	0.6	1.0	0.7	0.5	0

* Children in school are regarded as 0.1 units regardless of age.

0.2 consumer unit, medium-sized ones as 0.3 unit, and large animals as 0.5 unit.

The distribution of consumer-worker ratios for all married households was divided into five equal classes, and two married households were chosen from each class. The collection of data on time allocation (see Appendix C.3) necessitated the selection of households on the basis of their location within a "tolerable distance" from my residence because my two field assistants, who had to visit the households several times during each observation day, required supervision. Thus, it was impractical to use a completely random, stratified sample. However, because those who lived near my residence did not have peculiar characteristics that distinguished them from other villagers, no bias was introduced. Selected characteristics of the sample are listed in Table B.3.

Data were collected on the area cultivated for subsistence and coffee gardens, time allocated to various activities, consumption patterns, and income from coffee production and marketing. If one assumes that cattle bosses have a greater commitment to commercial endeavors than other Kapanarans do, an objection might be raised that the data on time allocation are biased toward greater inputs into income-producing endeavors than is charac-

TABLE B.3

Characteristics of the Sample of Households, July 1976

House-hold Number	Consumer-Worker Ratio	< 10		10-14		Unmarried Adult 15 +		Married Adult		55 +	
		M	F	M	F	M	F	M	F	M	F
1	1.60	2				1		1	1		
2*	1.84	1			1			1	2		
3*	1.45	2		1	1		1	1	3		
4*	2.63	1	2	1		1		1	1		
5*	1.68			1	1	2		1	1		
6*	1.67		1	1				1	1		
7	1.90	1		1		1		1	1		
8	1.10		1					1	1		
9	3.05		2					1	1		
10	0.95							1	1		
11										1	
12										1	
13											1

Total of 53 individuals.
* Household of cattle boss.

teristic of the village population generally because the number of cattle bosses in the sample is not proportional to their number in the total population. However, a comparison of the time spent by married adults in the cattle bosses' households with that spent by married adults in the other households in the sample on subsistence agriculture, coffee production, and cattle raising (see Table B.4) refutes such a possibility. The difference in the means for the percentage of time spent on subsistence agriculture is not statistically significant. The other means for coffee production and cattle husbandry were not tested because of the similarity in the results.

TABLE B.4

Percentage of Time Allocated to Productive Activities by Married Adults, Sample of Households

	Percentage of Time	
Activity	Households of Cattle Bosses	Households of Others
Subsistence Agriculture*	23.2	18.4
Coffee Production	7.1	7.5
Cattle Husbandry	2.1	2.0

* $t = 1.16 < 1.86 = t_{8DF}$; $p < .05$, 1 tail

Variance Ratio $= 2.55 < 6.39 = F_{4,4DF}$; $p < .05$

Estimation of Beer Consumption

From the end of June to December, I asked people every one to three days who had bought beer in the village and confirmed as many of the sales as possible with the purchasers. After short one- or two-week absences from the village, I also made the same enquiries, though the level of accuracy is slightly less for such periods because more purchases are likely to have been forgotten. Data from January to the end of June are estimates based on the rate of increase in monthly sales of beer from outlets in Kainantu. However, I presume that the rate at which purchases rose in this period in 1977 was more dramatic in Kapanara than in Kainantu because many civil servants who live in town have stable incomes which enable them to purchase beer throughout the year, whereas in the village purchases are more dependent on the seasonality of coffee income. Supporting this estimation, the data from Kainantu indicate that the rate at which purchases of beer fell after the peak of the coffee season was greater in the village than in town.

Methodology for the Collection
of Dietary Data

The collection of representative dietary data is extremely difficult and time-consuming (McArthur 1977). My efforts at weighing the amount of food eaten were not particularly successful. However, my attempt to obtain data on the frequency of foods eaten was much more fruitful. During each labor study day (see below), I asked mostly the adult members of as many of the households in the sample as possible what all their household members had eaten the preceding day and how the food was prepared. Data for the morning meal and for food consumed the rest of the day were recorded separately. Thus, two records or two "meals" were noted each day. By dividing the day into two periods, a more accurate representation of the frequency with which villagers consumed various foods is obtained. Needless to say, the data are most accurate for the adults and less so for their children.

Time-Allocation Study

Table C.1 details the average time each person spent per week during three different parts of the year on various activities; data are presented according to the age, sex, and marital status of the members of the sample. To clarify the meaning of the categories that are not self-explanatory, the types of activities included in the categories are:

FOOD GARDENS

Establishing: clearing, drainage ditch construction, tillage, burning vegetation, planting

Fencing: cutting wood for fencing, erecting fences, maintaining fences

Maintenance: weeding, repairing damage by pigs, breaking the soil crust, preserving planting material

Recultivation: tilling the soil during the final harvest, cutting and burning old sweet potato vines in preparation for replanting the plot

OTHER SUBSISTENCE TASKS

Pig husbandry: feeding pigs, checking newborn piglets, searching for them, other husbandry activities

Arboriculture: harvesting and cultivating *Pandanus* trees and *Areca* palm trees

Firewood: cutting and collecting firewood

Tools, equipment, and clothing manufacture: making arrows, net bags, and leaf mats, repairing clothing, sharpening axes

DIETARY

Dietary at social events: eating, waiting for food to cook, preparing food for cooking at social occasions

SOCIAL AND OTHER

Card playing: playing cards and watching others play cards

Beer-related: drinking and carrying beer

Singsing-related: preparing decorations for a *singsing* and attending one

General social activities: attending exchanges, ceremonies, and other social gatherings, visiting

CASH ACTIVITIES IN THE VILLAGE

Coffee: process: pulping, washing, drying coffee; maintenance: weeding and pruning

Cattle: maintaining the fences, herding cattle, constructing new fences, planting improved pasture

PWD work and other: working with the road crew, trade store-related activities, selling home-grown food to others in the village

GOVERNMENT-RELATED ACTIVITIES IN THE VILLAGE

Attending the village court or activities related to the *Eria Komuniti*, listening to government officials, building a school in the village, constructing a government-funded water pipeline

OUTSIDE VILLAGE

School-related: attending school and the annual school party, helping repair the school, attending a meeting for parents of school children

Other: all other reasons people were outside the village, such as marketing

To obtain the data, random spot-check observations of activities were made between 7 a.m. and 7 p.m. on 69 days. The conversion of the number of observations into hours was accomplished by dividing the number of observations for a particular activity by the total number of observations and multiplying the result by 84, the number of hours in the week in the sampling frame (see Grossman 1979, 1984).

TABLE C.1

Mean Time Allocation

(Hours per person per week)

	Widowed +50 M&F	Married M	F	Unmarried Adult 15+ M	F	10-14 M	F	< 10
Number of Individuals	3	10	14	5	2	4	6	13
Number of Observations	67	191	282	78	44	56	118	199
FOOD GARDENS								
Establishing	5.0	7.9	8.3	2.2	9.6	0	1.4	0
Fencing	1.3	0.9	0	1.1	0	0	0.7	0
Maintenance	10.0	2.6	2.1	0	0	0	1.4	0
Harvesting	3.8	1.3	10.1	2.2	1.9	7.5	5.7	0
Recultivation	1.3	0	2.4	0	0	0	1.4	0
Total	*21.4*	*12.7*	*22.9*	*5.5*	*11.5*	*7.5*	*10.6*	*0*
OTHER SUBSISTENCE TASKS								
Pig Husbandry	5.0	0.9	0.9	0	0	0	1.4	0
Hunting and Gathering	1.3	1.3	1.2	3.2	0	0	0.7	0
Arboriculture	1.3	3.5	2.7	3.2	3.8	7.5	2.1	0.8
Firewood	2.5	0.9	0.6	1.1	9.6	0	0.7	0
Tools, Equipment, and Clothing Manufacture	2.5	1.3	2.1	0	9.6	0	0.7	0
Housebuilding	3.8	3.5	1.8	3.2	0	1.5	2.1	0.8
Total	*16.4*	*11.4*	*9.3*	*10.7*	*23.0*	*9.0*	*7.7*	*1.6*
CHILDCARE *Total*	*0*	*0.4*	*2.4*	*0*	*1.9*	*1.5*	*0*	*0*
HEALTH-RELATED								
Grooming, Washing, and Cleaning	1.3	3.1	2.1	2.2	3.8	1.5	3.6	1.7
Sick	0	0.4	0.9	0	0	1.5	0	0
Total	*1.3*	*3.5*	*3.0*	*2.2*	*3.8*	*3.0*	*3.6*	*1.7*
DIETARY								
Eating	6.3	2.2	3.3	2.2	7.6	1.5	5.7	10.1
Waiting for Food to Cook	1.3	3.5	3.3	1.1	5.7	3.0	2.9	2.6
Food Preparation	6.3	1.3	5.1	0	7.6	3.0	5.0	2.5
Dietary at Social Events	3.8	2.6	5.7	2.2	0	3.0	1.4	2.1
Total	*17.7*	*9.6*	*17.4*	*5.5*	*20.9*	*10.5*	*15.0*	*17.3*

TABLE C.1 (cont.)

Mean Time Allocation

(Hours per person per week)

	FEBRUARY TO APRIL							
	Widowed +50 M&F	Married M	Married F	Unmarried Adult 15+ M	Unmarried Adult 15+ F	10-14 M	10-14 F	< 10
Number of Individuals	3	10	14	5	2	4	6	13
Number of Observations	67	191	282	78	44	56	118	199
SOCIAL AND OTHER								
Card Playing	0	10.1	3.0	8.6	1.9	3.0	1.4	0
Beer-Related	0	0	0	0	0	0	0	0
Singsing-Related	5.0	6.2	1.8	2.2	0	0	0.7	1.3
General Social Activities	3.8	3.5	5.4	4.3	3.8	1.5	2.1	3.4
Idle	11.3	8.8	6.9	17.2	9.6	25.5	10.7	54.0
Other	1.3	0	0	0	0	0	0	0
Total	21.4	28.6	17.1	32.3	15.3	30.0	14.9	58.7
CASH ACTIVITIES IN THE VILLAGE								
Coffee—picking	0	0.4	1.8	1.1	1.9	0	2.1	0
processing	1.3	0.4	0.6	0	0	0	0.7	0
maintenance	2.5	2.2	2.7	0	0	0	0	0.4
selling	0	0	0	0	0	0	0	0
Cattle	0	1.8	1.5	0	0	0	0	0.8
PWD Work and Other	0	2.2	0	2.2	0	0	0	0
Total	3.8	7.0	6.6	3.3	1.9	0	2.8	1.2
GOVERNMENT-RELATED ACTIVITIES IN THE VILLAGE *Total*	0	2.2	0.6	1.1	0	3.0	0	0.4
OUTSIDE VILLAGE								
School-Related	0	1.3	0.6	17.3	0	13.5	28.5	0
Wage Labor	0	0.4	0	0	0	0	0	0
Other	2.5	6.6	4.5	6.5	5.7	6.0	0.7	2.5
Total	2.5	8.3	5.1	23.8	5.7	19.5	29.2	2.5

TABLE C.1 (*cont.*)

Mean Time Allocation

(Hours per person per week)

	Widowed +50 M&F	Married M	Married F	Unmarried Adult 15+ M	Unmarried Adult 15+ F	10-14 M	10-14 F	< 10
				MAY TO AUGUST				
Number of Individuals	3	10	14	4	2	4	6	13
Number of Observations	75	243	328	107	35	66	136	192
FOOD GARDENS								
Establishing	4.5	2.4	1.8	0.8	0	0	0.6	0
Fencing	3.4	1.4	0.3	0	0	0	0	0
Maintenance	6.7	1.0	2.3	0.8	4.8	0	0.6	0
Harvesting	6.7	2.4	11.3	1.6	2.4	1.3	6.2	1.8
Recultivation	1.1	0.4	1.3	0	0	1.3	0	0.4
Total	*22.4*	*7.6*	*17.0*	*3.2*	*7.2*	*2.6*	*7.4*	*2.2*
OTHER SUBSISTENCE TASKS								
Pig Husbandry	2.2	2.8	1.5	0.8	0	0	0.6	0
Hunting and Gathering	1.1	0.7	0.8	0	2.4	1.3	1.2	0.4
Arboriculture	0	0.4	0.5	0	0	0	0	0
Firewood	2.2	1.7	1.0	0.8	2.4	1.3	1.2	0.4
Tools, Equipment, and Clothing Manufacture	3.4	0.4	1.0	0	2.4	0	0.6	0
Housebuilding	0	1.0	0.5	0	0	1.3	0	0
Total	*8.9*	*7.0*	*5.3*	*1.6*	*7.2*	*3.9*	*3.6*	*0.8*
CHILDCARE *Total*	*0*	*0.4*	*3.3*	*0*	*4.8*	*1.3*	*1.2*	*0*
HEALTH-RELATED								
Grooming, Washing, and Cleaning	2.2	0.7	2.8	2.4	2.4	0	2.5	1.8
Sick	4.5	1.0	1.0	0	0	0	0	0
Total	*6.7*	*1.7*	*3.8*	*2.4*	*2.4*	*0*	*2.5*	*1.8*
DIETARY								
Eating	0	2.4	2.1	2.4	2.4	2.6	2.5	4.8
Waiting for Food to Cook	2.2	3.1	4.1	0	0	5.1	0	1.8
Food Preparation	7.8	0	4.4	0	0	1.3	3.7	1.3
Dietary at Social Events	3.4	2.1	3.6	0	0	0	1.9	1.8
Total	*13.4*	*7.6*	*14.2*	*2.4*	*2.4*	*9.0*	*8.1*	*9.7*

TABLE C.1 *(cont.)*

Mean Time Allocation

(Hours per person per week)

	MAY TO AUGUST							
	Widowed +50	Married		Unmarried Adult 15+		10-14		< 10
	M&F	M	F	M	F	M	F	
Number of Individuals	3	10	14	4	2	4	6	13
Number of Observations	75	243	328	107	35	66	136	192
SOCIAL AND OTHER								
Card Playing	0	13.5	5.4	12.6	21.6	6.4	6.8	0.9
Beer-Related	0	5.2	0.5	0	0	0	0.6	0
Singsing-Related	3.4	2.1	2.1	0	0	0	1.2	0
General Social Activities	7.8	6.2	9.0	3.1	9.6	5.1	3.1	3.9
Idle	12.3	7.6	3.8	14.9	12.0	22.9	13.6	60.8
Other	0	0	0	0	0	1.3	0	0
Total	*23.5*	*34.6*	*20.8*	*30.6*	*43.2*	*35.7*	*25.3*	*65.6*
CASH ACTIVITIES IN THE VILLAGE								
Coffee—picking	2.2	4.5	6.4	1.6	0	3.8	4.3	0.9
processing	3.4	3.5	2.8	0	0	0	3.7	0.9
maintenance	0	0	1.0	0	0	0	1.2	0.4
selling	0	0.7	0.5	0	0	0	0	0
Cattle	0	2.4	2.6	0	0	0	0	0
PWD Work and Other	0	3.8	1.3	0	0	1.3	1.2	0
Total	*5.6*	*14.9*	*14.6*	*1.6*	*0*	*5.1*	*10.4*	*2.2*
GOVERNMENT-RELATED ACTIVITIES								
IN THE VILLAGE *Total*	*0*	*1.0*	*0*	*0*	*2.4*	*1.3*	*0*	*0*
OUTSIDE VILLAGE								
School-Related	0	0.4	0	18.8	0	19.1	18.5	0
Wage Labor	0	2.4	1.3	20.4	4.8	5.1	0	0
Other	3.4	6.6	3.8	3.1	9.6	1.3	6.8	1.8
Total	*3.4*	*9.4*	*5.1*	*42.3*	*14.4*	*25.5*	*25.3*	*1.8*

TABLE C.1 (*cont.*)

Mean Time Allocation

(Hours per person per week)

			SEPTEMBER TO DECEMBER				
	Widowed +50	Married		Unmarried Adult 15+		10-14		< 10
	M&F	M	F	M	F	M	F	
Number of Individuals	3	10	13	4	1	4	6	13
Number of Observations	74	227	293	77	5	58	122	189
FOOD GARDENS								
Establishing	3.4	13.0	14.3	2.2	*	2.9	2.8	0
Fencing	5.7	1.5	0	3.3		0	0	0
Maintenance	5.7	1.5	1.2	0		0	0.7	0
Harvesting	13.6	1.1	8.6	1.1		1.5	7.6	0.9
Recultivation	2.3	0	0.3	0		0	0	0
Total	*30.7*	*17.1*	*24.4*	*6.6*		*4.4*	*11.1*	*0.9*
OTHER SUBSISTENCE TASKS								
Pig Husbandry	3.4	0.7	2.6	0		0	0	0
Hunting and Gathering	0	1.9	0	2.2		4.3	1.4	0.4
Arboriculture	0	0.7	0	0		0	0	0
Firewood	5.7	1.9	0.6	0		0	0.7	0.4
Tools, Equipment, and Clothing Manufacture	2.3	1.9	0.6	1.1		2.9	0	0
Housebuilding	1.1	1.1	0	0		0	0	0
Total	*12.5*	*8.2*	*3.8*	*3.3*		*7.2*	*2.1*	*0.8*
CHILDCARE *Total*	*0*	*1.1*	*2.0*	*0*		*0*	*1.4*	*0*
HEALTH-RELATED								
Grooming, Washing, and Cleaning	2.3	0.7	1.4	2.2		0	2.8	2.2
Sick	1.1	0	3.7	0		0	1.4	0.4
Total	*3.4*	*0.7*	*5.1*	*2.2*		*0*	*4.2*	*2.6*
DIETARY								
Eating	2.3	3.7	3.4	2.2		2.9	3.4	4.4
Waiting for Food to Cook	0	1.1	0.6	1.1		0	0	0.9
Food Preparation	14.8	3.0	4.0	3.3		2.9	4.1	1.8
Dietary at Social Events	2.3	1.5	4.9	0		2.9	1.4	1.8
Total	*19.4*	*9.3*	*12.9*	*6.6*		*8.7*	*8.9*	*8.9*

TABLE C.1 (cont.)

Mean Time Allocation

(Hours per person per week)

	Widowed +50 M&F	Married M	Married F	Unmarried Adult 15+ M	Unmarried Adult 15+ F	10-14 M	10-14 F	< 10
				SEPTEMBER TO DECEMBER				
Number of Individuals	3	10	13	4	1	4	6	13
Number of Observations	74	227	293	77	5	58	122	189
SOCIAL AND OTHER								
Card Playing	0	1.9	1.7	8.7		2.9	2.1	0
Beer-Related	0	1.9	0.3	0		0	0	0
Singsing-Related	0	0	0	0		0	0	0
General Social Activities	11.4	7.4	8.3	9.8		4.3	6.2	5.3
Idle	4.5	8.9	9.8	15.3		30.4	16.5	60.4
Other	0	0.4	0.3	1.1		1.5	0.7	0.4
Total	*15.9*	*20.5*	*20.4*	*34.9*		*39.1*	*25.5*	*66.1*
CASH ACTIVITIES IN THE VILLAGE								
Coffee—picking	0	0.7	2.3	0		0	1.4	0.9
processing	1.1	2.2	1.2	0		1.5	0	0.5
maintenance	1.1	1.1	0.6	0		0	0	0.5
selling	0	0	0.3	0		0	0	0
Cattle	0	0.7	1.2	0		0	0	0
PWD Work and Other	0	4.1	0	0		0	0.7	0
Total	*2.2*	*8.8*	*5.6*	*0*		*1.5*	*2.1*	*1.9*
GOVERNMENT-RELATED ACTIVITIES								
IN THE VILLAGE *Total*	0	*3.3*	*1.7*	*1.1*		*0*	*0*	*0*
OUTSIDE VILLAGE								
School-Related	0	3.0	2.9	22.9		18.8	26.2	1.3
Wage Labor	0	0.4	0.3	1.1		0	0	0
Other	0	11.8	5.2	5.5		4.3	2.7	1.3
Total	*0*	*15.2*	*8.4*	*29.5*		*23.1*	*28.9*	*2.6*

* Too few observations to be meaningful.

Alcock, M. B. 1964. The Physiological Significance of Defoliation on the Subsequent Regrowth of Grass-Clover Mixtures and Cereals. In *Grazing in Terrestrial and Marine Environments*. D. J. Crisp, ed., pp. 25-41. Oxford: Blackwell Scientific Publications.

Anderson, D. 1977. *An Economic Survey of Smallholder Coffee Producers—1976*. Port Moresby: Department of Primary Industry, Papua New Guinea.

Anderson, J. L. 1963. Introduction of Ruminants to New Guinea. In *Symposium on the Impact of Man on Humid Tropics Vegetation*, Goroka 1960, pp. 170-173. Canberra: Government Printer.

Barnes, J. A. 1962. African Models in the New Guinea Highlands. *Man* 62: 5-9.

Barrie, J. W. 1956. Coffee in the Highlands. *Papua and New Guinea Agricultural Journal* 11: 1-29.

Bates, Robert H. 1981. *Markets and States in Tropical Africa: the Political Basis of Agricultural Policies*. Berkeley: University of California Press.

Bates, Robert H., and Michael F. Lofchie, eds. 1980. *Agricultural Development in Africa: Issues of Public Policy*. New York: Praeger.

Beckman, Bjorn 1981. Ghana, 1951-78: the Agrarian Basis of the Post-colonial State. In *Rural Development in Tropical Africa*. Judith Heyer *et al.*, eds., pp. 143-167. New York: St. Martin's Press.

Bennett, J. W. 1976. *The Ecological Transition: Cultural Anthropology and Human Adaptation*. New York: Pergamon Press.

Berndt, C. H. 1953. Socio-Cultural Change in the Eastern Central Highlands of New Guinea. *Southwestern Journal of Anthropology* 9: 112-138.

Bernstein, Henry 1979. African Peasantries: a Theoretical Framework. *Journal of Peasant Studies* 6: 421-443.

——— 1981. Notes on State and Peasantry: the Tanzanian Case. *Review of African Political Economy* 21: 44-62.

Berry, Roger 1978. Papua New Guinea Review. *Australian Quarterly* 50: 101-112.

Bishop, P., and A. Redish 1977. Ranching in Papua New Guinea. Paper delivered at the Executive Development Seminar, Livestock. Department of Primary Industry, Port Moresby.

Bonnemaison, J. 1978. Custom and Money: Integration or Breakdown in Melanesian Systems of Food Production. In *The Adaptation of Traditional Agriculture: Socioeconomic Problems of Urbanization.* E. K. Fisk, ed., pp. 25-45. Canberra: Australian National University Press.

Boserup, E. 1965. *The Conditions of Agricultural Growth: the Economics of Agrarian Change Under Population Pressure.* London: Allen and Unwin.

Boyd, David 1975. Crops, Kiaps, and Currency: Flexible Behavioral Strategies Among the Ilakia Awa of Papua New Guinea. Unpublished doctoral dissertation, University of California, Los Angeles.

——— 1981. Village Agriculture and Labor Migration: Interrelated Production Activities among the Ilakia Awa of Papua New Guinea. *American Ethnologist* 8: 74-93.

Boyon, J. 1957. *Naissance d'un état africain: le Ghana.* Paris: Armand Colin.

Brookfield, H. C. 1964. The Ecology of Highland Settlement: Some Suggestions. *American Anthropologist* 66 [4, 2, Special Publication: *New Guinea: the Central Highlands*]: 20-38.

——— 1966. But Where Do We Go from Here? In *An Integrated Approach to Nutrition and Society: the Case of the Chimbu.* E. H. Hipsley, ed., pp. 49-63. New Guinea Research Unit Bulletin, No. 9. Canberra: Australian National University.

——— 1968a. The Money that Grows on Trees. *Australian Geographical Studies* 6: 97-119.

——— 1968b. New Directions in the Study of Agricultural Systems in Tropical Areas. In *Evolution and Environment.* E. T. Drake, ed., pp. 413-439. New Haven: Yale University Press.

——— 1972. Intensification and Disintensification in Pacific Agriculture. *Pacific Viewpoint* 13: 30-48.

——— 1973. Full Circle in Chimbu: a Study of Trends and Cycles. In *The Pacific in Transition: Geographical Perspectives on Adaptation and Change.* Harold Brookfield, ed., pp. 127-160. New York: St. Martin's Press.

Brookfield, H. C., and P. Brown 1963. *Struggle for Land: Agriculture and Group Territories among the Chimbu of the New Guinea Highlands.* Melbourne: Oxford University Press.

Brookfield, H. C., and D. Hart 1966. *Rainfall in the Tropical*

Southwest Pacific. Department of Geography Publication G/ 3. Canberra: Australian National University.

Brookfield, H. C., with D. Hart 1971. *Melanesia: A Geographical Interpretation of an Island World*. London: Methuen.

Brown, Paula 1963. From Anarchy to Satrapy. *American Anthropologist* 65: 1-15.

———— 1972. *The Chimbu: a Study of Change in the New Guinea Highlands*. Cambridge: Schenkman.

Brown, Paula, and H. C. Brookfield 1967. Chimbu Settlement and Residence: a Study of Patterns, Trends and Idiosyncracy. *Pacific Viewpoint* 8: 119-151.

Bryceson, Deborah Fahy 1980. Changes in Peasant Food Production and Food Supply in Relation to the Historical Development of Commodity Production in Pre-Colonial and Colonial Tanganyika. *Journal of Peasant Studies* 7: 281-311.

Cairns, I. 1976. An Economic Analysis of the Papua New Guinea Beef Supply. Unpublished manuscript. Department of Primary Industry, Papua New Guinea.

Chappell, H. G., *et al.* 1971. The Effect of Trampling on a Chalk Grassland Ecosystem. *Journal of Applied Ecology* 8: 869-882.

Chayanov, A. V. 1966. *The Theory of Peasant Economy*. D. Thorner, B. Kerblay, and R.E.F. Smith, eds. Homewood: Richard D. Irwin.

Clark, R. D. 1970. Some Economic Aspects of Indigenous Cattle Development in the Eastern Highlands District of T.P.N.G. Paper delivered at the A.N.Z.A.A.S. Conference, Port Moresby.

Clarke, William C. 1966. From Extensive to Intensive Shifting Cultivation: a Succession from New Guinea. *Ethnology* 5: 347-359.

———— 1971. *Place and People: an Ecology of a New Guinea Community*. Berkeley: University of California Press.

———— 1973. The Dilemma of Development. In *The Pacific in Transition: Geographical Perspectives on Adaptation and Change*. H. C. Brookfield, ed., pp. 275-298. New York: St. Martin's Press.

Clarke, William C., and J. M. Street 1967. Soil Fertility and Cultivation Practices in New Guinea. *Journal of Tropical Geography* 24: 7-11.

Cliffe, Lionel 1976. Rural Political Economy of Africa. In *The Political Economy of Contemporary Africa*. Peter C. W. Gutkind and Immanuel Wallerstein, eds., pp. 112-130. Beverly Hills: Sage.

Cliffe, Lionel 1977. Rural Class Formation in East Africa. *Journal of Peasant Studies* 4: 195-224.

Commonwealth of Australia 1940a. *Territory of Papua Annual Report for the Year 1938-1939.* Canberra: Commonwealth Government Printer.

———— 1940b. *Report to The Council of the League of Nations on the Administration of the Territory of New Guinea for Year 1938-39.* Canberra: Commonwealth Government Printer.

———— 1952a. *Territory of Papua Annual Report for the Period 1st July, 1950, to 30th June, 1951.* Canberra: Commonwealth Government Printer.

———— 1952b. *Report to the General Assembly of the United Nations on the Administration of the Territory of New Guinea from 1st July, 1950, to 30th June, 1951.* Canberra: Commonwealth Government Printer.

———— 1953. *Territory of Papua Annual Report for the Period 1st July, 1951, to 30th June, 1952.* Canberra: Commonwealth Government Printer.

Connell, John 1978. *Taim Bilong Mani: the Evolution of Agriculture in a Solomon Island Society.* Canberra: Australian National University Press.

———— 1979. The Emergence of a Peasantry in Papua New Guinea. *Peasant Studies* 8: 103-137.

Cowen, Michael 1981. The Agrarian Problem: Notes on the Nairobi Discussion. *Review of African Political Economy* 20: 57-73.

Datoo, B. A. 1977. Peasant Agricultural Production in East Africa: the Nature and Consequences of Dependence. *Antipode* 9: 70-78.

Davidson, J. L. 1968. Growth of Grazed Plants. *Proceedings of the Australian Grasslands Conference*, Perth.

Deere, Carmen Diana, and Alain de Janvry 1979. A Conceptual Framework for the Empirical Analysis of Peasants. *American Journal of Agricultural Economics* 61: 601-611.

de Janvry, Alain 1981. *The Agrarian Question and Reformism in Latin America.* Baltimore: Johns Hopkins University Press.

Dejean, Eliane De Latour 1980. Shadows Nourished by the Sun: Rural Social Differentiation among the Mawri of Niger. In *Peasants in Africa: Historical and Contemporary Perspectives.* Martin A. Klein, ed., pp. 105-141. Beverly Hills: Sage.

Densley, D.R.J. n.d. Livestock. In *Agriculture in the Economy: a*

Series of Review Papers. Bob Densley, ed. Port Moresby: Department of Primary Industry, Papua New Guinea.

Dewey, Kathryn G. 1979. Agricultural Development, Diet and Nutrition. *Ecology of Food and Nutrition* 8: 265-273.

—— 1981. Nutritional Consequences of the Transformation from Subsistence to Commercial Agriculture in Tabasco, Mexico. *Human Ecology* 9: 151-187.

Duffey, E., *et al.* 1974. *Grassland Ecology and Wildlife Management.* London: Chapman and Hall.

Du Toit, B. M. 1974. *Akuna: A New Guinea Village Community.* Rotterdam: A. M. Balkema.

Dwyer, R.E.P. 1954. Coffee Cultivation in Papua and New Guinea. *Papua and New Guinea Agricultural Journal* 9: 1-5.

Eaton, Peter 1981. The Plantation Redistribution Scheme in Papua New Guinea. Paper delivered at the Waigani Seminar, Port Moresby.

Edmond, D. B. 1966. The Influence of Animal Treading on Pasture Growth. *Proceedings of the Tenth International Grasslands Congress,* Finland, pp. 453-458.

Elwert, Georg, and Diana Wong 1980. Subsistence Production and Commodity Production in the Third World. *Review* 3: 501-522.

Epstein, A. L. 1969. *Matupit: Land, Politics and Change among the Tolai of New Britain.* Canberra: Australian National University Press.

Feachem, Richard 1973. The Raiapu Enga Pig Herd. *Mankind* 9: 25-31.

Feil, Daryl K. 1982. From Pigs to Pearlshells: the Transformation of a New Guinea Highlands Exchange Economy. *American Ethnologist* 9: 291-306.

Finney, Ben R. 1965. *Polynesian Peasants and Proletarians.* Polynesian Society Reprints Series, No. 9. Wellington: The Polynesian Society.

—— 1973. *Big-Men and Business: Entrepreneurship and Economic Growth in the New Guinea Highlands.* Canberra: Australian National University Press.

Fisk, E. K. 1964. Planning in a Primitive Economy: from Pure Subsistence to the Production of a Market Surplus. *Economic Record* 40: 156-174.

—— 1971. Labour Absorption Capacity of Subsistence Agriculture. *Economic Record* 47: 366-378.

—— 1975. The Response of Nonmonetary Production Units to

Contact with the Exchange Economy. In *Agriculture in Development Theory*. Lloyd G. Reynolds, ed., pp. 53-83. New Haven: Yale University Press.

———— 1978. Traditional Agriculture and Urbanization: Policy and Practice. In *The Adaptation of Traditional Agriculture: Socioeconomic Problems of Urbanization*. E. K. Fisk, ed., pp. 345-377. Canberra: Australian National University Press.

Fitzpatrick, Peter 1980. *Law and State in Papua New Guinea*. London: Academic Press.

Flannery, K. V. 1968. Archeological Systems Theory and Early Mesoamerica. In *Anthropological Archeology in the Americas*. B. J. Meggers, ed., pp. 67-87. Washington, D. C.: Anthropological Society of Washington.

Fleckenstein, F. von 1975. Ketarovo: Case Study of a Cattle Project. In *Four Papers on the Papua New Guinea Cattle Industry*, pp. 91-138. New Guinea Research Bulletin, No. 63. Canberra: Australian National University Press.

Fleuret, Patrick and Anne Fleuret 1980. Nutrition, Consumption, and Agricultural Change. *Human Organization* 39: 250-260.

Ford, E. 1974. Climate. In *Papua New Guinea Resource Atlas*. E. Ford, ed., pp. 8-9. Milton: Jacaranda Press.

Forde, D., and M. Douglas 1956. Primitive Economics. In *Man, Culture and Society*. H. L. Shapiro, ed., pp. 330-344. New York: Oxford University Press.

Foster, George M. 1973. *Traditional Societies and Technological Change*, second edition. New York: Harper and Row.

Gerritsen, Rolf 1979. Groups, Classes and Peasant Politics in Ghana and Papua New Guinea. Unpublished doctoral dissertation, Australian National University.

Glick, L. B. 1972. Sangguma. In *Encyclopedia of Papua and New Guinea*. P. Ryan, ed., pp. 1029-1030. Melbourne: Melbourne University Press.

Golson, Jack 1981a. Agricultural Technology in New Guinea. In *A Time to Plant and a Time to Uproot*. Donald Denoon and Catherine Snowden, eds., pp. 43-54. Port Moresby: Institute of Papua New Guinea Studies.

———— 1981b. New Guinea Agricultural History: a Case Study. In *A Time to Plant and a Time to Uproot*. Donald Denoon and Catherine Snowden, eds., pp. 55-64. Port Moresby: Institute of Papua New Guinea Studies.

Good, Kenneth 1979. The Formation of the Peasantry. In *Development and Dependency: the Political Economy of Papua*

New Guinea. Azeem Amarshi, Kenneth Good, and Rex Mortimer, eds., pp. 101-122. Melbourne: Oxford University Press.

Good, Kenneth, and Mike Donaldson 1980. Development of Rural Capitalism in PNG: Coffee Production in the Eastern Highlands. Occasional Paper, No. 1. Port Moresby: Institute of Papua New Guinea Studies.

Graham, R. 1973. Results of Experimental Work to Date. Paper delivered at the Pasture Conference, Lae. Department of Agriculture, Stock and Fisheries, Papua New Guinea.

Gross, Daniel R., and Barbara Underwood 1971. Technological Change and Caloric Costs: Sisal Agriculture in Northeastern Brazil. *American Anthropologist* 73: 725-740.

Grossman, Larry 1979. Cash, Cattle and Coffee: the Cultural Ecology of Economic Development in the Highlands of Papua New Guinea. Unpublished doctoral dissertation, Australian National University.

——— 1980. The Beef Cattle Industry in Papua New Guinea: the Implications of Past Programmes for Future Planning. In *Cattle Ranches Are About People: Social Science Dimensions of a Commercial Feasibility Study.* Michael A.H.B. Walter, ed., pp. 17-42. Boroko: Institute of Applied Social and Economic Research.

——— 1982. Beer Drinking and Subsistence Production in a Highland Village. In *Through a Glass Darkly: Beer and Modernization in Papua New Guinea.* Mac Marshall, ed., pp. 59-72. Boroko: Institute of Applied Social and Economic Research.

——— 1983. Cattle, Rural Economic Differentiation, and Articulation in the Highlands of Papua New Guinea. *American Ethnologist* 10: 59-76.

——— 1984. Collecting Time-Use Data in Third World Rural Communities. *Professional Geographer* 36.

——— (In press). Sheep, Ceremonial Exchange, and Coffee in Papua New Guinea. *Geographical Review.*

Gudeman, Stephen 1978. *The Demise of a Rural Economy: from Subsistence to Capitalism in a Latin American Village.* London: Routledge and Kegan Paul.

Guillet, David 1981. Surplus Extraction, Risk Management and Economic Change Among Peruvian Peasants. *Journal of Development Studies* 18: 3-24.

Gunton, R. J. 1974. A Banker's Gamble. In *Problem of Choice: Land in Papua New Guinea's Future.* Peter G. Sack, ed., pp. 107-114. Canberra: Australian National University Press.

Gutteridge, M. 1977. Pasture Improvement in the Highlands. Paper delivered at the Executive Development Seminar, Livestock. Department of Primary Industry, Port Moresby.

Haantjens, H. A. 1970. Soils of the Goroka-Mount Hagen Area. In *Lands of the Goroka-Mount Hagen Area, Papua-New Guinea*. H. A. Haantjens, ed., pp. 80-103. Melbourne: Commonwealth Scientific and Industrial Research Organization.

Harding, Thomas G. 1972. Land Tenure. In *Encyclopedia of Papua and New Guinea*. P. Ryan, ed., pp. 604-610. Melbourne: Melbourne University Press.

Harris, G. T. 1974. Rural Business Development in the Koroba Sub-District of the Southern Highlands. Department of Economics Discussion Paper, No. 11. University of Papua New Guinea.

———— 1978. Cash Cropping or Subsistence Farming—a PNG Dilemma. *Pacific Islands Monthly* 49: 61-63.

Harriss, John 1982. *Capitalism and Peasant Farming: Agrarian Structure and Ideology in Northern Tamil Nadu*. Bombay: Oxford University Press.

Haswell, Margaret 1975. *The Nature of Poverty*. London: Macmillan.

Hays, T. E. 1974. Mauna: Explorations in Ndumbu Ethnobotany. Unpublished doctoral dissertation, University of Washington.

Henderson, P. M. 1972. A Sugar *Usina* in British Honduras. In *Technology and Social Change*. H. R. Bernard and P. J. Pelto, eds., pp. 136-163. New York: Macmillan.

Henty, E. E., and G. H. Pritchard 1975. *Weeds of New Guinea and Their Control*. Botany Bulletin, No. 7. Lae: Department of Forests.

Heyer, Judith, *et al.* 1981. Rural Development. In *Rural Development in Tropical Africa*. Judith Heyer *et al.*, eds., pp. 1-15. New York: St. Martin's Press.

Hide, R. 1974. On the Dynamics of Some New Guinea Highland Pig Cycles. Unpublished manuscript.

———— 1975. Coffee Production in Sinasina. Unpublished manuscript.

Hogbin, H. I. 1939. *Experiments in Civilization: the Effects of European Culture on a Native Community of the Solomon Islands*. London: Routledge and Kegan Paul.

Hogendorn, Jan S. 1975. Economic Initiative and African Cash Farming: Pre-Colonial Origins and Early Colonial Developments. In *Colonialism in Africa 1870-1960: the Economics of Colonialism*. Peter Duignan and L. H. Gann, eds., pp. 283-328. Cambridge: Cambridge University Press.

Holzknecht, H. A. 1974. *Anthropological Research and Associated Findings in the Markham Valley of Papua New Guinea*. Research Bulletin, No. 15. Port Moresby: Department of Agriculture, Stock and Fisheries.

Howard, Rhoda 1980. Formation and Stratification of the Peasantry in Colonial Ghana. *Journal of Peasant Studies* 8: 61-80.

Howlett, Diana R. 1962. A Decade of Change in the Goroka Valley, New Guinea: Land Use and Development in the 1950s. Unpublished doctoral dissertation, Australian National University.

———— 1973. Terminal Development: from Tribalism to Peasantry. In *The Pacific in Transition: Geographical Perspectives on Adaptation and Change*. Harold Brookfield, ed., pp. 249-273. New York: St. Martin's Press.

———— 1980. When Is a Peasant Not a Peasant: Rural Proletarianisation in Papua New Guinea. In *Time and Place*. J. N. Jennings and G.J.R. Linge, eds., pp. 193-210. Canberra: Australian National University Press.

Howlett, Diana R., *et al.* 1976. *Chimbu: Issues in Development*. Canberra: Australian National University Press.

Hughes, Ian 1978. Good Money and Bad: Inflation and Devaluation in the Colonial Process. *Mankind* 11: 308-318.

Hutton, Caroline, and Robin Cohen 1975. African Peasants and Resistance to Change: a Reconsideration of Sociological Approaches. In *Beyond the Sociology of Development: Economy and Society in Latin America and Africa*. Ivar Oxaal *et al.*, eds., pp. 105-130. London: Routledge and Kegan Paul.

Hyden, Goran 1980. *Beyond Ujamaa in Tanzania: Underdevelopment and an Uncaptured Peasantry*. Berkeley: University of California Press.

International Bank for Reconstruction and Development 1965. *The Economic Development of the Territory of Papua and New Guinea*. Baltimore: Johns Hopkins Press.

Jeffries, D. J. 1979. *From Kaukau to Coke: a Study of Rural and Urban Food Habits in Papua New Guinea*. Canberra: Australian National University.

Johnson, A. 1972. Individuality and Experimentation in Traditional Agriculture. *Human Ecology* 1: 149-159.

———— 1975. Time Allocation in a Machiguenga Community. *Ethnology* 14: 301-310.

Keesing, R. M. 1978. The Kwaio of Malaita: Old Values and New Discontents. In *The Adaptation of Traditional Agriculture: Socioeconomic Problems of Urbanization*. E. K. Fisk, ed., pp. 180-195. Canberra: Australian National University Press.

Kjekshus, Helge 1977. *Ecology Control and Economic Development in East African History*. Berkeley: University of California Press.

Klein, Martin A. 1980. Introduction. In *Peasants in Africa: Historical and Contemporary Perspectives*. Martin A. Klein, ed., pp. 9-43. Beverly Hills: Sage.

Kumar, A. and M. C. Joshi 1972. The Effects of Grazing on the Structure and Productivity of the Vegetation Near Pilani, Rajasthan, India. *Journal of Ecology* 60: 665-674.

Lacey, Roderic 1981. Agricultural Production on the Eve of Colonialism. In *A Time to Plant and a Time to Uproot*. Donald Denoon and Catherine Snowden, eds., pp. 65-84. Port Moresby: Institute of Papua New Guinea Studies.

Lam, N. V. 1979. Incidence of Agricultural Export Taxation in Papua New Guinea. *Journal of Development Studies* 15: 177-193.

———— 1982. A Note on the Nature and Extent of Subsistence Surplus in Papua New Guinea. *Pacific Viewpoint* 23: 173-185.

Langness, L. L. 1968. Bena Bena Political Organization. *Anthropological Forum* 2: 180-198.

———— 1972. Violence in the New Guinea Highlands. In *Collective Violence*. J. F. Short, Jr., and M. E. Wolfgang, eds., pp. 171-185. Chicago: Aldine.

———— 1975. The Nupasafa Cattle: Rural Development in the Eastern Highlands. In *Four Papers on the Papua New Guinea Cattle Industry*. New Guinea Research Bulletin, No. 63, pp. 67-90. Canberra: Australian National University Press.

Lappé, Frances Moore, and Joseph Collins 1977. *Food First: Beyond the Myth of Scarcity*. New York: Ballantine.

Laycock, D. C. 1972. Gambling. In *Encyclopedia of Papua and New Guinea*. P. Ryan, ed., pp. 475-478. Melbourne: Melbourne University Press.

Leche, T. F. 1977. Effects of a Sodium Supplement on Lactating Cows and their Calves on Tropical Native Pastures. *Papua New Guinea Agricultural Journal* 28: 11-17.

Lele, Uma 1981. Rural Africa: Modernization, Equity, and Long-Term Development. *Science* 211: 547-553.

Leopold, L. B. 1956. Land Use and Sediment Yield. In *Man's Role in Changing the Face of the Earth*. W. L. Thomas, Jr., ed., pp. 639-647. Chicago: University of Chicago Press.

Liddle, M. J., and K. G. Moore 1974. The Microclimate of Sand Dune Tracks: the Relative Contribution of Vegetation Removal and Soil Compression. *Journal of Applied Ecology* 11: 1057-1068.

Lingenfelter, Sherwood G. 1977. Socioeconomic Change in Oceania. *Oceania* 48: 102-120.

Lipton, Michael 1977. *Why Poor People Stay Poor: a Study of Urban Bias in World Development*. Cambridge: Harvard University Press.

Lockwood, Brian 1971. *Samoan Village Economy*. Melbourne: Oxford University Press.

Lofchie, Michael F. 1980. Introduction. In *Agricultural Development in Africa: Issues of Public Policy*. Robert H. Bates and Michael F. Lofchie, eds., pp. 100-112. New York: Praeger.

McAlpine, J. R. 1970. Climate of the Goroka-Mount Hagen Area. In *Lands of the Goroka-Mount Hagen Area, Papua-New Guinea*. H. A. Haantjens, ed., pp. 66-79. Melbourne: Commonwealth Scientific and Industrial Research Organization.

McAlpine, J. R., *et al.* 1975. *Climatic Tables for Papua New Guinea*. Melbourne: Commonwealth Scientific and Industrial Research Organization.

McArthur, M. 1977. Nutritional Research in Melanesia: a Second Look at the Tsembaga. In *Subsistence and Survival: Rural Ecology in the Pacific*. T. Bayliss-Smith and R. Feachem, eds., pp. 91-128. London: Academic Press.

McIntyre, D. S. 1974. Soil Sampling Techniques for Physical Measurements. In *Methods for Analysis of Irrigated Soils*. J. Loveday, ed., pp. 12-20. Commonwealth Bureau of Soils Technical Communication, No. 54. Farnham Royal: Commonwealth Agricultural Bureaux.

McKillop, R. F. 1965. Native Owned Cattle in the Eastern Highlands—a Review of Progress to Date. *Rural Digest* 7: 11-19.

———— 1975. Catching the Didiman. *Administration for Development* 3: 14-21.

———— 1976a. Helping the People in Papua New Guinea? A Case Study of a Cattle Introduction Programme. Paper presented at the Conference of the Sociological Association of Australia and New Zealand, La Trobe University.

——— 1976b. *A Brief History of Agricultural Extension in Papua New Guinea*. Extension Bulletin, No. 10. Port Moresby: Department of Primary Industry.

——— n.d. Problems of Access: Agricultural Extension in the Eastern Highlands of Papua New Guinea. Unpublished manuscript.

Mafeje, A., and Audrey I. Richards 1973. The Commercial Farmer and His Labour Supply. In *Subsistence to Commercial Farming in Present-Day Buganda*. Audrey I. Richards, Ford Sturrock, and Jean M. Fortt, eds., pp. 179-197. Cambridge: Cambridge University Press.

Malynicz, G. 1976. A Demographic Analysis of Village Pig Production. Paper delivered at the Waigani Seminar, Port Moresby.

Maruyama, M. 1963. The Second Cybernetics: Deviation-Amplifying Mutual Causal Processes. *American Scientist* 51: 164-179.

May, R. J., and Ronald Skeldon 1977. Internal Migration in Papua New Guinea: an Introduction to Its Description and Analysis. In *Change and Movement: Readings on Internal Migration in Papua New Guinea*. R. J. May, ed., pp. 1-26. Canberra: Australian National University Press.

Meeuwig, R. O., and P. E. Packer 1976. Erosion and Runoff on Forest and Range Lands. In *Watershed Management on Range and Forest Lands*. H. F. Heady *et al.*, eds., pp. 105-116. Logan: Utah State University.

Meggitt, M. J. 1958. The Enga of the New Guinea Highlands: Some Preliminary Observations. *Oceania* 28: 253-330.

——— 1971. From Tribesmen to Peasants: the Case of the Mae Enga of New Guinea. In *Anthropology in Oceania*. L. R. Hiatt and C. Jayawardena, eds., pp. 191-209. Sydney: Angus and Robertson.

Meier, Gerald M. 1975. External Trade and Internal Development. In *Colonialism in Africa 1870-1960: the Economics of Colonialism*. Peter Duignan and L. H. Gann, eds., pp. 427-469. Cambridge: Cambridge University Press.

Meillassoux, C. 1972. From Reproduction to Production: a Marxist Approach to Economic Anthropology. *Economy and Society* 1: 93-105.

Milford, R., and D. J. Minson 1966. The Feeding Value of Tropical

Pastures. In *Tropical Pastures*. W. C. Davies and C. L. Skid-more, eds., pp. 106-114. London: Faber and Faber.

Mitchell, D. D., II 1976. *Land and Agriculture in Nagovisi: Papua New Guinea*. Madang: Kristen Press.

Moulik, T. K. 1973. *Money, Motivation and Cash Cropping*. New Guinea Research Bulletin, No. 53. Canberra: Australian National University Press.

Munnull, J. P., and D.R.J. Densley n.d. Coffee. In *Agriculture in the Economy: a Series of Review Papers*. Bob Densley, ed. Port Moresby: Department of Primary Industry, Papua New Guinea.

Myint, H. 1958. The "Classical Theory" of International Trade and the Underdeveloped Countries. *Economic Journal* 68: 317-337.

———— 1969. International Trade and the Developing Countries. In *International Economic Relations*. Paul A. Samuelson, ed., pp. 15-35. London: Macmillan.

Newman, Philip L. 1965. *Knowing the Gururumba*. New York: Holt, Rinehart and Winston.

Nietschmann, Bernard 1973. *Between Land and Water: the Subsistence Ecology of the Miskito Indians, Eastern Nicaragua*. New York: Seminar Press.

———— 1979. Ecological Change, Inflation, and Migration in the Far West Caribbean. *The Geographical Review* 69: 1-24.

Nye, P. H., and D. J. Greenland 1960. *The Soil Under Shifting Cultivation*. Commonwealth Bureau of Soils Technical Communication, No. 51. Farnham Royal: Commonwealth Agricultural Bureaux.

Ogan, E. 1966. Drinking Behavior and Race Relations. *American Anthropologist* 68: 183-188.

Palmer, Robin, and Neil Parsons, eds. 1977. *The Roots of Rural Poverty in Central and Southern Africa*. Berkeley: University of California Press.

Papua New Guinea:

Bureau of Statistics 1970/71. *Rural Industries*. Port Moresby.
———— 1975/76. *Rural Industries*. Port Moresby.

Coffee Marketing Board 1972. *Coffee Growing in . . . Papua New Guinea*. Port Moresby: Department of Information and Extension Services.

Kainantu Bulumakau Association 1973. Minutes for Meeting on 27 June 1973. Department of Primary Industry, Kainantu. File 14.1.7(1).

Linsley, G. 1948/49. Kainantu Patrol Report, No. 7. Department of Native Affairs, Kainantu Sub-District.

National Statistical Office 1982. *Summary of Statistics, 1979.* Port Moresby.

—— n.d. 1980 National Population Census. Pre-Release: Summary of Final Figures. Port Moresby.

Papua New Guinea Development Bank 1974/75. *Annual Report and Financial Statement.* Port Moresby.

—— 1976/77. *Annual Report and Financial Statement.* Port Moresby.

—— 1977. *Interim Report and Financial Statements,* 1st July 1977-31st December 1977. Port Moresby.

Post-Courier (Papua New Guinea) 1979. Rice Restrictions a Worry. 10th April, p. 15.

Pataki-Schweizer, K. J. 1980. *A New Guinea Landscape: Community, Space, and Time in the Eastern Highlands.* Seattle: University of Washington Press.

Payer, Cheryl 1975. Coffee. In *Commodity Trade of the Third World.* Cheryl Payer, ed., pp. 154-168. London: Macmillan.

Porter, Philip W. 1979. *Food and Development in the Semi-Arid Zone of East Africa.* Syracuse: Maxwell School of Citizenship and Public Affairs.

—— 1981. Problems of Agro-Meteorological Modeling in Kenya. *Interciercia* 6: 226-233.

Powell, J. M. 1976. Ethnobotany. In *New Guinea Vegetation.* K. Paijmans, ed., pp. 106-183. Canberra: Australian National University Press.

Purdy, D. J. 1972. Cattle Industry. In *Encyclopedia of Papua and New Guinea.* P. Ryan, ed., pp. 137-141. Melbourne: Melbourne University Press.

Quartermain, Alan R. 1980. Livestock. In *South Pacific Agriculture: Choices and Constraints.* R. Gerard Ward and Andrew Proctor, eds., pp. 261-292. Canberra: Australian National University Press.

Radford, Robin 1972. Missionaries, Miners and Administrators in the Eastern Highlands. *Journal of the Papua and New Guinea Society* 6: 85-105.

Raikes, Philip 1978. Rural Differentiation and Class-Formation in Tanzania. *Journal of Peasant Studies* 5: 285-325.

Rappaport, R. A. 1968. *Pigs for the Ancestors: Ritual in the Ecology of a New Guinea People.* New Haven: Yale University Press.

———— 1971. The Flow of Energy in an Agricultural Society. *Scientific American* 224: 117-132.

—— 1977. Maladaptation in Social Systems. In *The Evolution of Social Systems*. J. Friedman and M. J. Rowlands, eds., pp. 49-71. London: Duckworth.

Robbins, R. G. 1963. The Anthropogenic Grasslands of Papua and New Guinea. In *Symposium on the Impact of Man on Humid Tropics Vegetation*, Goroka 1960, pp. 313-329. Canberra: Government Printer.

———— 1970. Vegetation of the Goroka-Mount Hagen Area. In *Lands of the Goroka-Mount Hagen Area, Papua-New Guinea*. H. A. Haantjens, ed., pp. 104-118. Melbourne: Commonwealth Scientific and Industrial Research Organization.

Roberts, R. J. 1978. The Theory and Use of Alternative Stocking Rates to Control Pasture Pests. Paper delivered at the Second Australian Conference of Grassland Invertebrate Ecology, Palmerston North.

Roseberry, William 1976. Rent, Differentiation, and the Development of Capitalism among Peasants. *American Anthropologist* 78: 45-58.

Russell, E. W. 1966. Soils and Soil Fertility. In *Tropical Pastures*. W. C. Davies and C. L. Skidmore, eds., pp. 30-45. London: Faber and Faber.

Sahlins, M. D. 1963. Poor Man, Rich Man, Big-Man, Chief: Political Types in Melanesia and Polynesia. *Comparative Studies in Society and History* 5: 285-303.

———— 1972. *Stone Age Economics*. Chicago: Aldine.

Salisbury, R. F. 1962. *From Stone to Steel: Economic Consequences of a Technological Change in New Guinea*. Melbourne: Cambridge University Press.

———— 1964. Changes in Land Use and Tenure Among the Siane of the New Guinea Highlands (1952-61). *Pacific Viewpoint* 5: 1-10.

Samoff, Joel 1980. Underdevelopment and Its Grass Roots in Africa. *Canadian Journal of African Studies* 14: 5-36.

Schindler, A. J. 1952. Land Use by Natives of Aiyura Village, Central Highlands, New Guinea. *South Pacific* 6: 302-307.

Schottler, J. 1977. The Feeding Value of Lowland Native and Semi-Improved Pastures in the Lowlands. Paper delivered at the Executive Development Seminar, Livestock. Department of Primary Industry, Port Moresby.

Scott, G.A.J. 1974. Effects of Shifting Cultivation in the Gran

Pajonal, Eastern Peru. *Proceedings, Association of American Geographers* 6: 58-61.

Scott, James C. 1976. *The Moral Economy of the Peasant: Rebellion and Subsistence in Southeast Asia.* New Haven: Yale University Press.

Sexton, L. 1982. New Beer in Old Bottles: an Innovative Community Club and Politics as Usual in the Eastern Highlands. In *Through a Glass Darkly: Beer and Modernization in Papua New Guinea.* Mac Marshall, ed., pp. 105-118. Boroko: Institute of Applied Social and Economic Research.

Shanin, Teodor 1971. Introduction. In *Peasants and Peasant Societies.* Teodor Shanin, ed., pp. 11-19. Harmondsworth: Penguin Books.

———— 1979. Defining Peasants: Conceptualizations and De-Conceptualizations Old and New in a Marxist Debate. *Peasant Studies* 8: 38-60.

Shenton, R. W., and Louise Lennihan 1981. Capital and Class: Peasant Differentiation in Northern Nigeria. *Journal of Peasant Studies* 9: 47-70.

Shenton, R. W., and Mike Watts 1979. Capitalism and Hunger in Northern Nigeria. *Review of African Political Economy* 15: 53-62.

Sillitoe, Paul 1979. Stone Versus Steel. *Mankind* 12: 151-161.

Sorenson, E. Richard 1972. Socio-Ecological Change among the Fore of New Guinea. *Current Anthropology* 13: 349-372.

Standish, Bill 1978. The Big-man Model Reconsidered: Power and Stratification in Chimbu. IASER Discussion Paper, No. 22. Boroko: Institute of Applied Social and Economic Research.

Stavenhagen, Rodolfo 1975. *Social Classes in Agrarian Societies.* Garden City: Anchor Press.

———— 1978. Capitalism and the Peasantry in Mexico. *Latin American Perspectives* 5: 27-37.

Steensberg, Axel 1980. *New Guinea Gardens: a Study of Husbandry with Parallels in Prehistoric Europe.* London: Academic Press.

Strathern, Andrew J. 1969. Finance and Production: Two Strategies in New Guinea Highlands Exchange Systems. *Oceania* 40: 42-67.

———— 1982a. The Division of Labor and Processes of Social Change in Mount Hagen. *American Ethnologist* 9: 307-319.

———— 1982b. Tribesmen or Peasants. In *Inequality in New Guinea*

Highlands Societies. A. Strathern, ed., pp. 137-157. Cambridge: Cambridge University Press.

Taussig, M. 1978. Peasant Economics and the Development of Capitalist Agriculture in the Cauca Valley, Colombia. *Latin American Perspectives* 5: 62-90.

Thomas, A. S. 1960. The Tramping Animal. *British Grassland Society Journal* 15: 89-93.

Tosh, John 1978. Lango Agriculture During the Early Colonial Period: Land and Labour in a Cash-Crop Economy. *Journal of African History* 19: 415-439.

Townsend, D. 1977. The 1976 Coffee Boom in Papua New Guinea. *Australian Geographer* 13: 419-422.

Townsend, W. H. 1969. Stone and Steel Tool Use in a New Guinea Society. *Ethnology* 8: 199-205.

UNESCO/UNFPA 1977. *Population, Resources and Development in the Eastern Islands of Fiji: Information for Decision-Making.* General Report No. 1 of the UNESCO/UNFPA Population and Environment Project in the Eastern Islands of Fiji. Canberra: Australian National University.

Uyassi, M. 1975. Improving Access: the Komuniti Kaunsils. *Yagl-Ambu* 2: 51-64.

Vayda, A. P., A. Leeds and D. B. Smith 1961. The Place of Pigs in Melanesian Subsistence. In *Proceedings of the American Ethnological Society.* V. E. Garfield, ed., pp. 66-77. Seattle: University of Washington Press.

Vergopoulos, Kostas 1978. Capitalism and Peasant Productivity. *Journal of Peasant Studies* 5: 446-465.

Vickery, P. J. 1972. Grazing and Net Primary Production of a Temperate Grassland. *Journal of Applied Ecology* 9: 307-314.

Waddell, Eric 1972. *The Mound Builders: Agricultural Practices, Environment, and Society in the Central Highlands of New Guinea.* Seattle: University of Washington Press.

———— 1975. How the Enga Cope with Frost: Responses to Climatic Perturbations in the Central Highlands of New Guinea. *Human Ecology* 3: 249-273.

Ward, Alan 1981. Customary Land, Land Registration and Social Equality. In *A Time to Plant and a Time to Uproot.* Donald Denoon and Catherine Snowden, eds., pp. 249-264. Port Moresby: Institute of Papua New Guinea Studies.

Ward, R. Gerard, *et al.* 1974. *Growth Centres and Area Improvement in the Eastern Highlands District.* Canberra: Department of Human Geography, Australian National University.

Ward, R. Gerard, and Epeli Hau'ofa 1980. The Demographic and Dietary Contexts. In *South Pacific Agriculture: Choices and Constraints*. R. Gerard Ward and Andrew Proctor, eds., pp. 27-48. Canberra: Australian National University Press.

Watson, J. B. 1967. Tairora: the Politics of Despotism in a Small Society. *Anthropological Forum* 2: 53-104.

——— 1970. Society as Organized Flow: the Tairora Case. *Southwestern Journal of Anthropology* 26: 107-124.

Watson, J. B., and V. Watson 1972. *Batainabura of New Guinea*. Ethnocentrism Series. New Haven: Human Relations Area Files.

Watts, Michael J. 1979. The Etiology of Hunger: the Evolution of Famine in a Sudano-Sahelian Region. *Mass Emergencies* 4: 95-104.

Westermark, George D. 1982. "Old Talk Dies Slowly": Land Mediation in Agarabi. Paper delivered at the American Anthropological Association Meeting, Los Angeles.

Wharton, Clifton R., Jr. 1971. Risk, Uncertainty, and the Subsistence Farmer. In *Economic Development and Social Change: the Modernization of Village Communities*. George Dalton, ed., pp. 566-574. Garden City: Natural History Press.

Williams, Gavin 1976. Taking the Part of Peasants: Rural Development in Nigeria and Tanzania. In *The Political Economy of Contemporary Africa*. Peter C. W. Gutkind and Immanuel Wallerstein, eds., pp. 131-154. Beverly Hills: Sage.

——— 1978. Imperialism and Development: a Critique. *World Development* 6: 925-936.

Wisner, B. 1977. Man-Made Famine in Eastern Kenya: the Interrelationship of Environment and Development. In *Landuse and Development*. P. O'Keefe and B. Wisner, eds., pp. 194-215. London: International African Institute.

Wolf, Eric R. 1955. Types of Latin American Peasantry: a Preliminary Discussion. *American Anthropologist* 57: 452-471.

——— 1966. *Peasants*. Englewood Cliffs, N.J.: Prentice-Hall.

——— 1969. *Peasant Wars of the Twentieth Century*. New York: Harper and Row.

Wurm, S. A., ed. 1978. Language Maps of the Highlands Provinces, Papua New Guinea. Pacific Linguistics, Series D, No. 11. Canberra: Department of Linguistics, Australian National University.

Young, E. 1973. *The People of the Upper Ramu: a Socio-Demographic Survey of Agarabi-Gadsup*. Department of Geography Occasional Paper, No. 8. Port Moresby: University of Papua New Guinea.

——— 1977. Population Mobility in Agarabi/Gadsup, Eastern Highlands Province. In *Change and Movement: Readings on Internal Migration in Papua New Guinea*. R. J. May, ed., pp. 173-202. Canberra: Australian National University Press.

Young, Michael 1971. *Fighting with Food*. Cambridge: Cambridge University Press.

Library of Congress Cataloging in Publication Data

Grossman, Lawrence S., 1948-
    Peasants, subsistence ecology, and development in the highlands
of Papua New Guinea.

    Bibliography: p.
    Includes index.
    1. Peasantry—Papua New Guinea. 2. Rural development—
Papua New Guinea. 3. Agriculture—Economic aspects—
Papua New Guinea. 4. Human ecology—Papua New Guinea.
I. Title.
HD1537.P26G76 1984   305.5′63   84-42581
ISBN 0-691-09406-3 (alk. paper)

Lawrence S. Grossman is Assistant Professor of Geography at
Virginia Polytechnic Institute and State University.